# OUT OF THE DESERT

## A STORY OF PALESTINE, PLOESTI AND BEYOND

JOHN E. BLUNDELL

**Turner Publishing Company**
412 Broadway • P.O. Box 3101
Paducah, Kentucky 42002-3101
(270) 443-0121

Copyright © 1999 John E. Blundell
Publishing Rights Turner Publishing Company

Turner Publishing Company Staff:
Editor: Bill Schiller
Designer: Herbert C. Banks II

Library of Congress Card Catalog No: 99-67807
ISBN: 1-56311-536-0

This is a Limited Edition
Printed in the USA

# TABLE OF CONTENTS

## DEDICATION

This book is dedicated to my children:

ANDY
LINDA
SCOTT
Who insisted that I write down my story.

And to BRIAN
Our grandson, for whom the book was written.

And to my dear wife DOROTHY
Without whose encouraging words this might never have happened.

# INTRODUCTION

by
Sid McMath
Major General, US Marine Corps Reserve
World War II Veteran

John Blundell's *Out of the Desert* bring us a personal account of an air mission critical to the war against the Nazis during WWII that is not generally known or appreciated. He has told a story that needs to be told. Few people know about the oil fields of Ploesti and its strategic significance in the winning of the war.

The Ploesti oil refineries supplied 90% of the oil that fueled the Nazi war machine. The refineries were ringed with anti-aircraft guns, protected by squadrons of fighter planes, and equipped with radar to warn of an impending attack.

Based in Lete, Libya, after five months of intensive preparation and training, the 98th Bomber Group, 9th Air Force, launched an attack against the Ploesti refineries. The Bomb Group approached the target at a low level in an effort to avoid detection by the enemy radar. The assault wave striking its target ignited the massive tanks. Planes following the assault wave were engulfed in flames. This greatly enhanced the hazards to the planes and crew, which they were experiencing from withering anti-aircraft fire.

Mission accomplished. A critical source of fuel for the German forces was severely damaged—but at what cost: fifty B-24 bombers had been launched by the 98th Bomb Group for the attack; only ten with their crews returned to base that night.

Colonel John R. Kane, Group Commander, fully recognizing the significance of the mission and the price to be paid, was determined to have photographic evidence as to the results of the attack. He assigned John Blundell, chief of the Bomber Photographic Section, to improvise and install in each plane the necessary photographic equipment, aligned and positioned to clearly record the attack's impact on the refineries.

Colonel Kane had informed the pilots and crew that some would not return from this mission. He directed that they arrange their personal effects and write to members of their families, to be mailed if they were a casualty.

Sergeant Blundell was assigned the additional duty of helping to arrange for the transfer of personal effects of the pilots and crew members who were lost in this mission.

A letter from a 22 year-old airman to his parents reveals the courage, the unselfish devotion to duty, and the nobility of the young men who fought in the Second World War and won a victory against tyranny for freedom in the world.

John Blundell's personal account of his experiences is good reading and a significant contribution to the history of the GI's who served in World War II.


Sid McMath

*Sid McMath was Governor of the State of Arkansas, 1949-1953. He gave John Blundell his first job after returning from service as the State of Arkansas Official Photographer.*

# CHAPTER ONE

## *The Long Trip*

My overseas journey in World War II with the United States Army Air Corps had a most unusual beginning. The Air Corps reserved space for us on a French luxury ship (the *Louis Pasteur*, captained by L.A. Fraser) which had been converted to a troop ship used to carry prisoners of war to America and Canada and return loaded with American GI's. What a way to go to war! We were told this was one of the seven largest ships in the world, and the fourth fastest. How ironic! The Air Force song, "Off We Go Into The Wild Blue Yonder," was never intended to describe the way we went to war. We went by train at night from Ft. Dix to the dock where we loaded the ship. I entered the ship that was to be our home for the next 32 days with all my personal belongings in a canvas bag thrown over my shoulder. The Sergeant told us to get rid of all that we could and to take only the essentials. We were told that new supplies would be issued when we arrived at our destination. My essentials included only one change of clothing and three boxer shorts. I did travel light as they had suggested, but later on I wished I had put in a few more clothes. Doing my own laundry kept me busy nearly every day, but fortunately with the wind and sun on the ship, there was no drying problem. I also took some shaving items, a toothbrush and a Bible. For me, the Bible was the most essential. Nothing else really mattered to me.

As a youth, I was raised by my mother and father to believe the Bible was a guideline for our daily living. I was told it was a resource book for whatever problems we encountered. By the time I was 15 years old, I was teaching a Sunday School to a group of my peers. The Bible was important to me, and my mother told me that I needed to read it every day. She set the example for me, for I saw her studying and reading it every day. I have fond memories of this.

We departed from New York Harbor early in the morning of July 15, 1942, just as all New York was busy going to work. The morning sun filtered through the early fog to highlight that wonderful skyline to us. This gave me a strong desire to return some day and discover what was going on behind this enchanting scene. As we passed the Statue of Liberty, what a thrill I received at seeing this gallant Lady. I was seeing her for the first time on a troop ship bound for where, I did not know. She had always been a symbol of hope for immigrants entering this country for the first time. Then she was a symbol to me of all I was leaving behind. A great lump filled my throat. The Lady took on a new meaning. Standing there in all her majesty, somehow I knew, more than ever, that the light she held must never go out. Each of us on that ship had a great sense of national pride. We were called by our country to give sacrificially all that we could. Each of us was willing to do so.

As our ship slipped silently past the statue I thought about my high school days in Arkansas. In 1938 I attended Little Rock High School where there were 4,200 enrolled students. 563 seniors graduated with me that year. Members of both the local and national press at that time were anxious to know if graduating seniors would go to war for their country. Hitler was experiencing great success in Europe, and they wanted to know what the youth thought about his armies rolling through Europe. Would we be willing to fight in this war? Reporters came to the school campus to interview some of the graduating seniors. Since I was President of the student body, they sought me out for an interview.

Members of the press gathered on the steps at the front of the school for good photo opportunities. The interview began with a statement: "This seems to be a fun-loving generation. You dance the 'Big Apple,' you seek pleasures and don't seem to be worried about anything."

The "Big Apple" was a popular dance of that era that was sweeping the country and young people in particular loved it. It was the craze of the times, and we had fun with it even though the adults thought we had lost our minds with such a silly dance. First, you would form a big circle, then take a step toward the center and kick your right leg as high as you could. Next, you would take two steps backward and kick your left leg as high as you could. You would repeat this over and over. As the music played, you would snap your fingers and clap your hands while keeping up with the music. Any number of people could participate, and you didn't need an equal number of boys and girls. It may have looked crazy but it sure was fun.

Then the members of the press asked the big question: "All America wants to know, are the youth of this country willing to go to war?"

I replied, "It is our feeling that this is Europe's war and they need to settle their own problems."

"Are you saying the American youth will not fight?" one of them asked. I felt that was the answer they wanted to hear.

"No, that's not at all what I am saying. This country was founded on strong moral principles and a deep belief in God. This feeling has been handed down through the generations, and it is still a viable part of the young people of today. True, we do believe in having fun, but when it comes to defending our country, you don't need to worry about this generation. We are a generation that went through the great depression together and we learned from that experience how much we need each other. If we are to make it; we will make it together. Should

this Country be attacked or drawn into the war, you will see young people united and ready to go to war without hesitation."

Now, two years later, I was on a troop ship sailing off to war, and my thoughts were drawn back to the interview on the steps of Little Rock High School. There we were exactly seven months and seven days after Pearl Harbor on a ship loaded with 5,000 of America's finest youth, 18 to 24 years of age, sailing off to war with countless others to follow. I wondered what "All America" wanted to know now. What was the press thinking and writing about today? I hoped the doubts were now gone, and the answer clear. If America had only known its own youth there would never have been any doubt.

On the afternoon of our first day out I was comfortably settled on the sports deck reading Stephen Crane's classic, "Red Badge of Courage," which I had picked up at a USO Canteen. Suddenly an explosion was heard off the port side of our ship. It was so close and so large that even this great ship shook with the concussion. As we left the harbor our escort, a destroyer was soon dropping a depth charge at the sighting of a sub. I thought to myself, "They are getting in some good practice. I hope they will be there when we need them." So I quickly found my place and went back to reading. This book was more interesting than the "war" at hand. As I continued to read, two more depth charges were released, and our ship shook again. I began to hear excitement from the other soldiers as they ran to the port side to see what was going on. This didn't bother me because I thought I had read about how they would practice firing their big guns and dropping depth charges once at sea to make sure everything was working okay.

Man, was I wrong. When two airplanes showed up and entered into the battle I knew this was no practice. The airplanes circled low over us like birds of prey circle a field on a hot summer day. Later, I was told there were two enemy submarines awaiting our departure from the harbor in order to follow us. One airplane dropped three bombs and the destroyer dropped two more depth charges. Later, one of the gunners on our ship told me that one sub was totally destroyed, and the other was badly damaged. They could tell this by the oil and all the debris that came floating to the surface.

What a close call for us. Two subs destroyed while I in innocence read my book. We could have been lost at sea before we ever got into the war. Fortunately the enemy subs never got a shot off at us. Could this be a good omen that we would indeed return to walk the streets of New York another day? Perhaps so. That was our dream, and we would hold on to it. Time would tell.

There were about five thousand troops aboard this French ship, or so the ship's crew told us. This was not official because troop movement in those days was top secret. For security purposes we loaded the ship at night. As soon as we got aboard we were assigned bunk space for sleeping and eating. It was all the same space. At night you hung up a hammock above the table where you took your two meals each day. My space was four levels below deck. It was always hot and the food odors were so bad that I didn't spend much time below. The food was not very good and it was always too close and stuffy in there to eat. By the time I went down four decks below I had lost all desire to eat. As long as I had money, I bought Hershey bars at the canteen. I lived on Hershey bars for several days. I'm not sure whether the ship ran out of Hershey bars or I ran out of money first, but that wonderful pleasure came to an end. I didn't get much food or sleep, but chose to sleep on deck with many others rather than be stuck four decks below with a hammock in the hole. In fact I suspect we all felt this way, but there wasn't enough room for all of us to be on the top deck. Each night there would be a mad scramble, like musical chairs, to stake out a place for the human body, not necessarily to stretch out, but just big enough to lie down. Everywhere I turned I saw men sleeping on rafts, on sheds, on racks, on floors, on stairs, or anywhere they could find to be in the fresh air. Some had blankets and duffel bags for pillows while others had nothing. They formed a huddled mass coiled in all shapes for sleeping as if they had been used to sleeping that way all of their lives.

How flexible were these American boys. I just knew that not one of these GI's had ever experienced anything quite like this but no one complained. This was the hand that had been dealt them and they made the best they could out of a bad situation. Seeing this great mass of human bodies you could just about believe all five thousand of these GI's must have been on the top deck. This gave me a feeling that nothing can beat the spirit of these men. There was a sense of deep pride and a strong love for our country shown in their attitude.

There wasn't much to do on ship except read and walk. I continued to read my book until I finished it and then I found someone who would trade books with me. This way I had a book to read most of the way over. When I wasn't reading, I walked around the ship for exercise, but also to see whom else I might run into for engaging conversation. After all we were all in the same boat, literally and figuratively, so why not make the best of it. I met a friend that liked to walk, so to get in a little exercise every day we would meet at the same place and the same time so we each had someone to walk with. We tried jogging but there were too many bodies around to do this. It turned out that walking was a good way to kill time. Before the boat trip was over, we figured we had walked all the way to Egypt.

After we had been at sea for sometime, and I had finished the book I was reading, I did a lot of deck walking. The weather was perfect. There was always a beautiful sky, plenty of sunshine, and a nice cool breeze blowing that somehow seemed to sweep away the cobwebs from my mind. As I walked along the deck, I would stop along the way and visit with a few GIs before going on to another group to see what was happening in their life. We did a lot of talking about the things that were important to us at this time. All of the things that were on the minds of a group of teenagers would be up for discussion. The girls we had known and the cars we had owned. We talked about the women we left behind, some were wives and others were girlfriends.; the family and fiends that were important to us; the cars we owned and the ones we didn't own but pretended we did, just to make conversation. The home life we had, whether good or bad, was always on our minds.

Often we spoke with the men on board the ship about the role we would play in the drama of a combat zone. It gave us something to do during the long and monotonous hours which seemed to roll on endlessly. We speculated about what it would be like, and wondered how we would it into the scheme of things. This we knew; it wouldn't be like it was in the states. The food would be different, the freedom we had would be gone, and we would live in much greater restrictions and tension. What would the accommodations be like, we wondered. What about the airfield? What would it be like and would there be quarters for all the different sections. I thought about the Photo Lab and hoped we would have a trailer equipped with everything we needed. It gave us something to think and wonder about, but of course we could only speculate about what it would be like and what was ahead for us. The uncertainties we didn't know about, but there was one thing for sure that we did know. Whatever the situation was, whatever conditions we would have to face; we would be ready and adapt to meet what was ahead for us. The way all the men had to adapt to this troop ship, affirmed this for me. The way they adapted the lousy food and the way they adapted to sleeping on deck, I knew we would be ready for what was ahead, for surely they could cope with anything.

These are the things we wondered and talked about on our way to war. The attitude of every one of us was to get over there and get it over with, so that we could return to our former life we had known was gone forever. We would change, life itself would change, and times and conditions would change. We would never again be the same. This is a lesson we had yet to learn.

One bright sunny day, I had been walking alone some time and stopped to look over the side of the ship for awhile. I took off my shirt because the warm sun and wind felt good blowing so hard on my body. I kept watching the water roll over and over. It seemed like perpetual motion. It never ended. I thought about what power it must take to propel this massive ship at such a fast pace through the waters. The sound and motion kept beating through my mind. I began to think about my situation and how I got here, and what the future would hold for me. My thoughts soon went back to January 14, 1941. I was working in downtown Little Rock, and just finished having lunch with my first cousin, Bill Roseberry. Bill lived with my parents and me while working in Little Rock. We said, "Goodbye. Will see you at supper time" and parted, each going our own separate ways. Later when I walked into the Federal Building where the Draft Board was located, I was shocked to see Bill standing in line to volunteer for the Army Air Corps. We were as close as two brothers could be, but neither of us had said a word about joining the Air Force. I got in the line behind him and received a serial number, only one digit higher than his.

Our country was not yet in war, but we were required to register for the draft. We both had a high draft number and would not have been called for another year or two, but the Draft Board had announced a new program that would allow men to volunteer for one year of service, fulfill our military obligation, and not be subject to be drafted later. We both signed up for one year and spent this time training for photo reconnaissance. We had almost served our time when the Japanese attack came at Pearl Harbor on December 7, 1941, drawing the United States into World War II. Because of this Bill and I were split up and assigned to different squadrons. My one-year enlistment turned into four and a half years of service, with three of those years in a combat area.

Meanwhile, to avoid a torpedo from

an enemy sub, our ship would change its course every six minutes. We were told it took seven minutes for a sub to lock in on our bearing and fire a torpedo. We therefore zigzagged all the way across the Atlantic Ocean. After nine days at sea we anchored just off the coast of Freetown, Sierra Leone, South Africa to take on fresh water and fuel oil.

Although we were offshore and not in a harbor, as soon as the natives saw the ship set anchors, the young men and boys paddled out in small one man boats to greet us. It was quite a distance out to deep water where we were, but about 25 of them paddled so fast it looked as if they were racing to get to us. They were so happy to see us and really put on a show.

Moving around the base of the ship so quickly, they reminded me of a school of fish, darting in and out and churning up the water. They would shout out, "Hey Joe, throw me something!" We were all "Joe" to them. No matter what you would throw to them they would let it fall into the water and then dive in to retrieve it. We were enjoying them so much everybody seemed to join in throwing something over and watching the native's dive for the object. One of the most popular items was a cigarette. They would come up with one in their mouth and then eat it, paper and all! But the guys keep throwing things such as T-shirts, socks, and money just to watch the rivalry. It was a contest for them to see who would get the item first. One of the GI's threw in a wristwatch. I don't know if the watch worked or not, but it sure created lots of excitement and made one of the natives very happy. After retrieving it in the water, one of the young men held the watch up for all to see that he was the winner of this great prize. He then put it on his arm as if he were wearing a fine piece of jewelry. The wet gold watch and its gold band glistened in the sunlight on his black wrist. You would think he had just found the "crown jewels," he was so proud! Hail to the new King. Hail to the new Chief. With his arm still held high, he waved goodbye to all of the "Joes" on the ship as he paddled back towards the shore. These natives retrieved lots of "loot" from the water, but they gave all of the "Joes" on the ship a good time and were very entertaining. Then they disappeared back to the shore where they came from as rapidly as they had appeared. Soon after that incident we set out to sea and on to Durban, South Africa.

On July 29, 1942 we crossed the equator, and the French crew had quite a surprise for us. Since the ship was originally built for pleasure cruises, the crew kept the equipment and costumes traditionally used in the ceremony of crossing the equator. According to Roman mythology, Neptune was the name of the Greek Sea God who controlled the seas. It would be unacceptable and unthinkable for this French crew to cross the equator without showing respect to King Neptune. Otherwise the God of the Sea might not look with favor upon their ship in his seas.

Members of the crew put on an impressive show about this important event. Of course it was all done in good fun. Nobody believed it but everybody enjoyed it. This broke up the monotony of the trip, and provided relief from tension.

When we arrived at Durban, South Africa on August 4, there was a group of South African soldiers singing for us as we pulled into the harbor. I honestly don't think I have ever heard singing with such beautiful rhythm and harmony. It sounded as if they had been singing like this for a long time. Many of their songs were familiar tunes we knew, American folk songs but with their own words. One song that they sang for us was to the tune of "Hand Me Down My Walking Cane." Of course those were not the words they were singing, but we recognized the tune. They sang with much joy and enthusiasm. It was so moving you could just feel the spirit of their songs. The GI's on the ship were touched by their music and responded by singing a song to them before we unloaded from our ship. It was all spontaneous. The men just erupted in song. Although I'm sure we didn't sound as good as they did, they responded with great enthusiasm. Then we joined together, both GI's and natives in singing "My Old Kentucky Home." They sang with broken English, but the feeling and meaning were strong. Spirits were lifted by this evidence of a bond of brotherhood between us. This was entirely unexpected, but oh, what a pleasure. It left us with a feeling of gratitude and a sense of thankfulness, thankful for something close to all of us. How this came about and who directed it was never clear to any of us. Let's say it just happened. A boat loaded with sad hearts had just been warmed for the next part of our journey.

We stayed in port for two days and were allowed to go into Durban both days, returning to the ship at night. It was a welcome relief for us. The food on ship was mainly fish or mutton, and we were ready for some different food.

The city was one of great contrast. The population was European and native Zulu, both modern and primitive; it was clean almost to the point of sterile and yet part of the city was very dirty, covered in filth and flies. I never had seen so many things painted white. The fences were white and even the rocks were painted white. The clean areas were very clean and the dirty areas were very dirty. These contrasts were a new experience for us. In amazement we saw how these people lived, how they ate, and how they worked in this mixed environment, and we gave thanks for being born in America. We enjoyed their hospitality, and their food, we saw many new and strange sights which I'm sure we will remember for a long time. For a young Arkansas boy who had not seen much of the World, this was exciting, enlightening and I might add quite unexpected.

In addition to the supplies we took on at Durban, we also picked up a small cadre of 250 Zulu solders. They were trained in their native ability to fight but somewhat disciplined by the British who trained them. They were dressed in British uniforms with each man carrying a rifle and a long knife, similar to what we know as a Bowie knife, in a scabbard strapped around their waist. They looked fierce as they boarded the ship. We were told they were experts in hand combat

*Cargo boats along the Suez Canal. Man pulls boat along with a harness attached to his shoulders.*

and were ferocious warriors who liked to fight at night. Once their knife was drawn from its scabbard, they were not permitted to put it back until that knife had drawn blood; either from a foe or from themselves. Believe me, they didn't look like they were interested in drawing their own blood! You could tell that was not what they had in mind. You would not want to have an encounter with them. One good look and you were thankful they were on our side. Later when we were in combat in North Africa, they were there for us when we needed them. We would realize just how fortunate we were that we had picked them up in Durban, SA. and they were part of our group.

Back out to sea. We started the last lap of our voyage, and perhaps the most dangerous part of our trip. Now we are to sail through the Indian Ocean, which was heavily infested with enemy submarines. For eleven more days and nights we zigzagged through these hostile waters. We depended on the speed of our ship to get us safely through this part of our journey. Once our ship had to make a very sharp change in direction to miss a torpedo. Another close call. By now we were all alone without an escort. We had an escort only for the first day at sea. The rest of the trip we were on our own and defenseless. Thankfully the torpedo missed us. It was late in the afternoon and I thought we were going to be washed over by a big wave of water when our ship made such a sharp turn.

My favorite spot for sleeping on deck was beside a large crate about half the size of a boxcar mounted on the aft of the ship. I liked this place because it knocked off some of the wind at night. Each day I would try to get there before dark to claim this spot. The crate was covered by a huge tarp and we could not see what was underneath. After many days someone got curious enough to investigate as to what was inside this crate. A triangular slit about three inches long was made in the tarp covering the crate. It was just large enough for a fist to go inside and explore inside the tarp. Much to our surprise this large crate was loaded

with onions! We didn't know if they were Vidalia Onions from the State of Georgia, but they were big, juicy, sweet onions. We didn't know why they were there or where they were going but why question it? They were there we were hungry and this was edible food. At a time when we really needed some nourishment, the Good Lord had provided and we were thankful! To say the food on our ship was not good would be an understatement. It was so terrible we ate very little of it. Our ship hauled prisoners of war, mostly Germans and Italians, from Europe to Canada and the United States. We felt like they were feeding us the same food they served to the POW's. Well, at least it was not the kind of food we were accustomed to and we stayed hungry the long trip.

We kept our little secret about the onions from others on ship, but each night thereafter, we would peel and eat onions like apples. I'm sure we all had onion breath, but with the sea air, and our stomachs delightfully full, who cared.

Finally after 32 days at sea, we landed at Port Said on the Suez Canal. Through some rough and calm waters; through close calls and near disaster, we rode the ocean waves before we finally came to our destination in the Middle East. At last, we were on firm ground. There we unloaded under cover of night at Port Said on the Suez Canal. We were taken to an Egyptian encampment where, still in darkness, we were fed a meal of camel stew (at least that is what we called it) and watermelon. We were both surprised and delighted to find watermelon in Egypt. For an old southern boy the watermelon was wonderful but not that stew. I wasn't too sure what was in it and didn't eat very much. We were also introduced to pita bread, which we loved. WOW, what a feast, pita bread and watermelon! It was very late at night, well after midnight, when we finished unloading and eating. For the rest of the night we slept out under the stars with a blanket and our duffel bag for a pillow. This ended day thirty-two of our long trip overseas. As the night rolled on, we slept very little and wondered, what is next for us?

# CHAPTER TWO

## *Getting It All Together*

The soldiers unloaded from our ship were scattered throughout the area around Egypt. We landed in a combat zone, so for our safety most troops were quickly dispersed. My group of GI's who were to join the aircrews of the 98th Bomb Group were left behind for three days. We were left to sleep in the sands of the desert near the Suez Canal before we went on to our destination. We began to wonder if anyone knew where we were or what to do with us. Finally, the Limeys came and picked us up in their Lorries (translation: The British Army came and picked us up in their trucks). They transported us to a rail-head where we caught an antique narrow-gauge train that took us to Palestine. It was a long, hard day's trip without food. The only redeeming factor of the trip was seeing many strange and interesting sights along the route. So far we had gone to war for the Air Corps by ship, by truck and now by a narrow-gauge train. We began to wonder where are the planes?

Only a few hundred of us who came over on the ship were moved north into Palestine (now known as Israel) where we were to join our aircrew. Some of us were based at St Jean, Palestine, while the others were stationed at Ramat David, Palestine. Both bases were near the city of Haifa. Although the war had not spread into this part of the Middle East, we traveled much of the time by night as a precaution against surprise attacks. A little after midnight, we arrived at our destination. We camped in an olive garden, not far from Beirut. Even after a long hard trip, this was an exciting time for me. What a thrill it was just being here in the Holy Land, in an olive garden. I was so grateful that

we had finally reached our destination safely, I dropped to my knees and thanked God. Afterwards, I spread my GI blanket on the soft ground and had a good night's sleep. Now safely on land at our destination, and without the constant motion of the ship, I felt secure at last and thought I would be able to sleep throughout the night. This excitement soon wore off since we had only our two blankets to sleep on the ground or concrete floors for three months before we got any cots. I tried both and found with only two blankets, the concrete was too hard to sleep on.

At last we had found our air crews, and were they glad to see us! Now we were in this safer environment it was time for us to get our air and ground crews together as a full working Squadron. Preparations for war must go on. Our planes and combat crews had arrived at our destination about three weeks ahead of the ground crews. By the time we arrived, they had already flown nine missions with the RAF crews servicing the planes and loading the bombs for them.

We were young kids from America's main streets, rural farms, and schools and colleges. Most of us were still green and had not experienced life. Here we were in a foreign land with no way to keep us in touch with the civilian life we had so abruptly left behind. We had not yet learned about fighting, drinking liquor, loose women, or shooting craps. We had a lot of tough lessons in life to learn. We were brought together in the hopes that we would bond together and build into the greatest force for democracy the world had seen. Yet untried and unchallenged, we had

*Forward minefields near Tobruk, Egypt*

much to prove to ourselves and to our country. The experiences we will yet encounter, will be the measure of men. Time would tell.

Since we were the first Americans to land in this part of the world, we had no American supplies or equipment. At first we had to live on British rations and later were issued British clothing and equipment.

At the time our group entered into this war, the situation in the Middle East had become critical. The German General Rommel and the British Field Marshall Montgomery were fighting across North Africa for control of this territory. Rommel at this time was winning decisively. The Port of Tobruk had just fallen to the Germans and Montgomery had to fall back to El Alamein, only 60 miles west of Alexandria. This was about 200 miles from Cairo, which made it even more critical. If Cairo fell to the Axis, all of the Middle East would be lost. The way the German and the

Italian armies were rolling, another 200 miles was nothing for them. They just stopped at Tobruk to catch their breath, regroup, and get the petrol they needed, then roll on again. They had traveled so far and so fast; they had run out of gasoline and water. The Allied Forces knew Rommel and his troops must be stopped while they regrouped.

We were told when leaving the States that our squadron would join the 98th Bomb Group to be stationed in India. This all changed when our B-24 planes stopped in Cairo to refuel. Winston Churchill persuaded President Roosevelt to allow our planes to fly bombing missions in support of General Montgomery. We were to strengthen his position by stopping the flow of supplies to Rommel's North Afrika Korps.

The Germans and Italians controlled the Mediterranean Sea with their ships and submarines. This gave them a tremendous advantage because their ships brought in all the equipment, fuel, oil, and other supplies they needed. Our mission was to cut off the free flow of goods to the enemy in North Africa. We were to bomb all the ports of North Africa, Sicily, Italy, and any ships at sea.

We were not prepared or equipped for this type of operation. However, our planes did start dropping bombs on the enemy tankers and ships wherever we found them, at sea or in the harbors. We continued doing this to help the Allies hold on at Tobruk. Montgomery and his Allied troops dug in and were able to hold off the enemy. While waiting on our regular supplies and equipment to come from the States, we flew missions daily. During this time, we were still flying out of Rama David, Palestine, and some of our other planes were at St. Jean. These were makeshift airfields with dirt runways, which had been developed in hurry and with a minimum of effort. With a Caterpillar No. 12 Motor Grader, painted olive drab, they could scrape off five or six inches of top soil, and you had an instant airfield. It is difficult to believe the adverse conditions under which our heavy bombers were operating from these strips, but with American ingenuity and willpower at work, our bombing missions were very effective. These strikes began to slow the enemy down and have a lasting effect on their momentum. Now the German Afrika Korps were experiencing a shortage of supplies and were unable to get all of the oil and fuel they required.

My assignment with the 98th Bomb Group, 343rd Squadron, was to set up a photo reconnaissance unit to take pictures and identify what our planes were hitting on bombing missions. This was no problem for me as this is what I had been trained to do while assigned to the 154th Air National Guard Reconnaissance Squadron. My major problem was the lack of equipment. The trailer the Air Corps had equipped as a Photo Lab was lost at sea. I was brought into this new Bomb Group to set up a Photo Section, now I had to find a solution to the problem of setting up a Photo Lab without the trailer and all of the related equipment that came with it. What a challenge! I had to find an answer.

Until I could get this Photo Section set up I was assigned to the Armament Section loading our planes with 2,000 lb. British bombs. We were having trouble mounting the bombs because they didn't exactly line up with the shackles on the B-24's bomb racks. One day as I was helping to load these bombs one of them came loose in the bomb bay of the plane and started falling. One of my good friends, Tatum Hendricks, and I were working together. As the bomb came towards us, I instinctively lunged at him knocking both of us away, clear of the bomb. There was no fear of the bomb exploding as the fuse had not yet been installed, but the sheer weight of the bomb would have crushed Tatum to death because his head was directly under the falling bomb.

As the bomb fell it hit Tatum a glancing blow on the side of his head. Tatum liked to brag about being the only person in the army hit in the head by a 2,000-lb. bomb and live to tell about it.

That did it for me. I went directly to the Commanding Officer of the 98th Bomb Group, Col. John R. "Killer" Kane. A very determined man with strong opinions. Col. Kane was a big Texan who earned his nickname, "Killer," for driving his men so hard. This was his style of leadership, very firm, but also very fair. He was a hard-driving commander but he never asked his men to do anything he would not do himself. Kane led his men by his example. One day we were standing in line for noon chow when Captain Ashcraft walked up and took his place at the front of the line. Col. Kane, who was in line with about a half dozen officers behind a number of enlisted men, waited until Ashcraft got in the front of the line before calling him down. Kane told him he was no better than the rest of the men and to take his place at the end of the line. This was the kind of leader Kane was and the men respected him for his fairness.

It was very presumptive of me, a Sergeant, to be telling him anything, but I did. I told him that I thought it was time for us to get our photo lab together and operational.

His cold steely eyes, looked directly at me as he said in very firm tones, "Sergeant, we didn't come over here to take pictures. We came here to fight a War."

With that attitude I was devastated but not defeated. None of us had ever thought we might lose our Photo Lab equipment before we received it.

I explained to the Col. that taking pictures was a vital part of this war. How else could he be certain of the success of the bombing strikes? Pictures could confirm what had been done or what was missed on the mission. Not only had I been trained to take and process reconnaissance photographs, but also I had been trained to interpret those photographs.

I said, "Colonel, I have ten experienced airmen in the photo section who are anxious to do the job they were sent over here to do. Right now our Photo Lab is not here and these men are just filling in the squadron wherever they can, looking for something to do. Sir, all I'm asking is a chance to prove my point. Give me the authority to set up a Lab here on the base, and I will show you how helpful bomb strike photos will be to you."

He could hear the intensity of my voice and recognize my desire to make this work. He said to me, "Okay, Sergeant, I will neither approve nor will I stop you from doing whatever it takes to prove your point. If you want to go out and round up the equipment and supplies to do this you will have to be personally responsible. If you think you

*Photo Section, 98th bomb Group.*

can accomplish this, go for it. And I'll be watching for the results."

That was all I needed. We started immediately to round up essential items from the British and from the nearby town of Haifa. As time moved on and with that good old American ingenuity we continued to work at putting together an operational photo lab. I began to look around and see what makeshift arrangements we could use to get started. I drew up a plan to take over a small building on the site that was used by the local people as a laundry. It would be necessary to turn this into a photo lab for processing and printing our bomb strike pictures. Most of our work would be done at night, so adapting this building was not too much of a problem. We requisitioned six aerial cameras from the British and got them with no trouble. With my rank of Staff Sergeant, my signature was all it took to get the equipment. Soon we were in business. We equipped six planes with cameras for their next missions.

All of this would have been easy with the photo trailer since it was completely equipped for field operation. I was told one of these trailers was on the way to us but a submarine in the Indian Ocean sank the ship carrying it. We continued our make-do operation for another three months hoping all the time that our lab and equipment would finally arrive. Still it did not come, but this didn't stop us. To keep our lab in operation we had to scrounge around and "borrow" from the British whatever we could find. Even under these difficult conditions we were getting good bomb strike pictures. This pleased Col. Kane so much he gave me access to a Jeep and authority to find whatever I needed from whatever source and from wherever I could find them. "Just keep these bomb strike photos coming in," he directed.

We continued to get more cameras from the British and our staff worked hard at doing a better job. Now working together the men assigned to the Photo section began to know each other better and learn what each one could

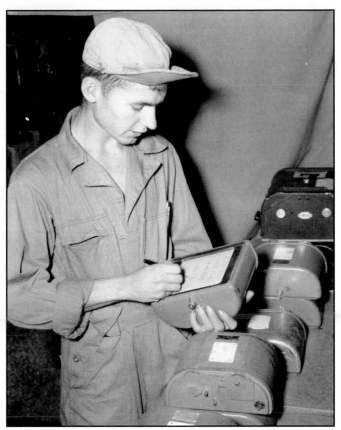

*Sgt. A.V. O'Neil checking out film magazines for the aerial cameras before mission.*

do. Our section began to develop into a strong unit. Our men were from all over the United States and had a diverse range of experience, capabilities and personalities. Two were from New Orleans, Sgt. Herman Willem and Cpl. Charles Fisher. Willem was our Cajun with a strong outgoing personality who fell in love soon after he arrived here. Fisher, in contrast to Willem, was very quiet and did all of the hand lettering on our bomb strike photos. Three were from the New England States, Sgt. Ed Whewell. Cpl. Hank Hallock, and Pvt. Bud Winston. All three of these men were excellent technicians and good soldiers. Cpl. John Marcinko was from McKees Rocks, Pennsylvania, and was as solid as a rock. Cpl. A.V. O'Neal came from South Carolina. We never found out what the A.V. stood for so we just called him "Avon." Cpl. Linderson was from Connecticut and did much of the film processing for us. Sgt. Eugene Gambrel hailed from Drew, Mississippi, so we called him "the drip from Drew." He was a happy fellow who took this in a

*John Marcinko looking at ruins of German aircraft.*

*Cpl. Hank Hallock drying his aerial film after processing.*

good-natured way and always wore a smile on his face. Cpl. Tatum Hendricks and I were both from Little Rock. Hendricks, whom we called "Doc," was probably the youngest man in our section, and one of the toughest. He would tackle anything, and was afraid of nothing. The Photo Officer assigned to our unit was 2nd Lt. Donald Browne. He was really a nice young man from Detroit, but didn't know the first thing about photo reconnaissance; we got along with him very well and called him "Brownie." I liked to kid him a lot and often told him, "Stick with me Brownie and I will make something out of you yet." He got several promotions while we were there and he would always come to me and ask me to pin the new rank on his uniform. I always considered this a real compliment. This was our start up crew.

My army fatigues had my initials, "JEB," stenciled on them. One day "Doc" Hendricks noticed what they spelled and said, "One of my favorite Generals in the Army was JEB Stuart. He was a Confederate General and the South's most brilliant Cavalry officer. He was a good man and the name fits you. From now on I think we should call you 'JEB.'" That name sort of stuck with me. For the rest of my service life, I was JEB.

We were eleven men who didn't know each other, but were brought together by the military because of our MSO (Military Service Occupation). For the unknown future we were to work together, eat together and sleep together, twenty four hours a day. We would come together as family, knowing each other and helping each other in all the ways we could. We were yet untested, but soon would find out just how strong we were.

To make our lab operational, I continued to get more and more of our supplies and equipment from the British. I procured so much equipment with no more than my signature my men began to worry about me. They said, "Sergeant, one of these days after the war is over someone from the British Government will come to Little Rock and present you with a bill for all of this stuff you are getting."

I said, "That is okay, I'll just ask them to put it on our 'Lend Lease' account and that should take care of that."

We didn't let that stop us. We never looked back, and the British never stopped filling my requests. I must tell you more about one of our men, Sgt. Willem, who fell in love with one of the local girls as soon as we hit Palestine. I kept hoping this was an infatuation that would soon pass, but it grew into full courtship and an engagement. Of course he could not get married overseas without permission from the Air Force. We tried to discourage this but were unsuccessful. I knew Sgt. Willem was of the Catholic faith, and his girl, Magda, was Jewish. With this background, we felt it would be difficult to make this marriage work. Her cultural background was so different from his American culture, it would be difficult for each of them to adapt to each other. Another obstacle they would have to face was being separated for a long and indefinite period of time, as he would be in service overseas and she would be in the States.

Sgt. Willem filled out the papers to get married, turned them in and waited for approval. The officer in charge let the papers lay around as long as he could before approving them and sending them on.

It is hard to imagine the full meaning of having access to a Jeep, being stationed in the Holy Land with a camera, and an assignment to go wherever needed to procure supplies. What discipline this demanded! True, there was a great temptation for a Bible student like myself to see the Holy Land as I would like to do. But yet, I knew we were sent here for a much more important task. With restraint, I continued working on building the kind of lab we needed. Sight seeing would have to come later.

As the days passed, we discovered the loss of our photo equipment was not our only loss. Many more of our supplies were lost at sea. In fact since we landed we had not received any American food, clothing or equipment. Not only had we not received our needed supplies, but also we had not been paid for several months. We had only the possessions we brought with us. We were able to get food and other supplies from the British, but mutton, lamb, tea, and the like, were not the kind of food I was accustomed to eat, nor what I wanted. We wore the British clothes okay, but we struggled with their food. We kept thinking we would get our American rations soon but they never came. Because of this I realized how fortunate we were that our troop ship had made it through the water safely.

The day we were issued British Army clothing was a real blow for us. How could this happen? We wondered if we would ever get anything more from the States. Even "K" rations and "C" rations would have tasted good to us by now. The entire 98th Bomb Group for the first year felt detached from anything American. We were with a large contingent of British soldiers, fighting what we felt was a British war, eating British rations, and wearing parts of British uniforms. We felt alone and abandoned. Mail from home was difficult to get through, and news from the US was little or none. We felt our own Air Force has completely forgotten us. We began to call ourselves "The Lost Air Force."

We were now three months out of the states and still no mail from home. We had not been paid since we left. We didn't even know what was going on back home. In fact we didn't know very much about what was going on in the rest of the world except for our own little world and the people around us. I guess this was the beginning of the "feel sorry for me" time. We went off to war in the Air Force on a French Ocean liner, and now we are wearing British uniforms and eating British food. We are in a world of strange people and disturbing events. There is no word from home and all we know about are the people around us. Slowly they became "family" because they were the only real people we knew. You eat, sleep and breathe the same air with them 24 hours a day, day after day and they become a part of you. We learned to care about each other and how to help each other. We are family. We have bonded.

The one thing that kept me going was my strong faith in God. When I left home I took my Bible with me, and I never let it get very far from me. Reading the Bible kept my spirits up. I was one of the few that had a Bible and often was asked by other GI's about my faith in God. Some of the fellows started coming to me with their problems. I always listened to them, and sometimes I would counsel with them and offer suggestions. It seemed as though I had become the unofficial chaplain for our outfit. I don't know how this got started, but I was a little older than most of the men in our group, and perhaps a bit more experienced. The guys started saying, "I need to come by and have you punch my TLC (Tough Luck Card) tonight, if that is okay."

What they really were saying was they needed to talk out some problem they had. They were in need of someone who would listen to them. This I could do, and I was glad to do so. Our organization was one of the first units shipped out and we did not have an opportunity to come together as a working unit before leaving the United States. Most of the men left so quickly they didn't have an opportunity to get all of their personal problems resolved. They brought these problems with them and they seemed to compound and get larger.

I had been asked to discourage Sgt. Willem about his desire to get married in Palestine to a Jewish girl. Try as hard as I could I did not get very far. He was still determined to marry his Jewish sweetheart.

Private Bud had another kind of a problem. He had lived a sheltered life and never been away from home. He had a father who was domineering and a mother who never crossed him, all of this being very hard on the boy. Bud was a high school dropout and lied about his age to get in the Air Force. He was not well adjusted and needed a lot

of guidance. When offered something, candy, cigarettes, food or drink, he would always take it, even if he didn't want it or could use it. He never refused anything. The kid didn't smoke and cigarettes were rationed and hard to come by, but just let someone offer him a cigarette, and Bud would take one. He would take the cigarette and toss it in his footlocker already chucked full of old dry cigarettes. Needless to say, this type of hoarding did not make Bud popular with anyone in the outfit. I asked him one day why he did that and he said he didn't know when he might start smoking or might need them.

Well at last our bombing missions were showing some effect. They had been so successful the Allies had dug in and were holding back Rommel's forces at Tobruk. It was like a lull before the storm. You felt both sides were regrouping and getting ready for a new offensive.

With our photo lab now working well, and our planes still making bomb strikes at the enemy supply lines, I felt it was time to move out and experience some of the sights in this Holy Land. With a few of my friends and Jim Bray who was a freelance journalist and photographer who had come to our Bomb Group, we took off in a Jeep on a three day pass to see the wonderful, exciting and historic city of Jerusalem.

*98th Bomb Group Photo Section, set up and operational.*

# CHAPTER THREE

## Palestine: The Land and The People

With things in good order at the Base, and a lull just before the big thrust, I felt I now had the time to investigate this land and its people. My desire was to see the country side, talk to individuals, visit historical sites, and learn more of their history and better understand the people who lived and worked in this land. I had read about this land in the Bible, and it intrigued me most of my life. Now was the time to see for myself and to experience first hand the land and its people. I received my first paycheck since leaving the States and now had money to spend. On the 16th of September 1942, I set off for Jerusalem. I did need to find some "hard to get" photo supplies, and where better to look than this great historic city. Checking out an Army Jeep from the car-pool, I took off with three of my buddies on a much anticipated journey.

We drove down the Mediterranean coast enjoying the fantastic blue waters of the Mediterranean Sea before turning inland towards the City of Jerusalem. My mind whirled with the thoughts of things I had been told about the Holy Land all through my formative years. To be here now presented a thrill for me. This is, after all, the "Promised Land," a good and broad land, a land flowing with milk and honey that God promised Moses and the Israelites. The phrase "flowing with milk and honey" simply was an expression meaning the land was fruitful and productive.

The land of Palestine has always been more significant than its size. Although relatively small, it has been the center of controversy and war with the great powers of ancient and modern times. Strategically located at the joining of Africa and Southwest Asia, Palestine has often been called "A Bridge" and "The Crossroads for the Nations." Geopolitical and historically, Palestine was linked to three major religions: Judaism, Christianity, and Islam. The city of Jerusalem was especially significant to these three religions because each considers the city holy.

Palestine was originally called Canaan and was settled by Canaanite tribes who controlled the area for more than a thousand years. In about 1500 BC Hebrew tribes entered the area. Later the Greek people, known as Philistines, populated this same area and it was from them that the name Palestine was derived. The Hebrew Army came in conflict with the Philistines and successfully defeated them. After this defeat the Hebrews established the kingdom of Israel. This flourished for about 75 years and then it split into two weaker kingdoms that fell under more powerful neighbors. At the time of Jesus the region had become a part of the Roman Empire, and was controlled by the Romans.

During World War I the British forces took command of this region. Following the war, The League of Nations granted Britain the right to administer Palestine under a League mandate committing the country of Palestine for future self-rule. Under the Balfour Declaration of November 1917, the British helped the growth of the Jewish population in Palestine through immigration so that the Jews would gain a majority and thereby gain self-rule.

Most Jewish immigrants came from Eastern Europe until 1930 when Jewish people from Germany started to enter Palestine in large numbers. After Adolf Hitler came to power in Germany, he persecuted the Jews and many were killed, but also many fled to Palestine. This sharp increase in Jewish immigration in the mid 1930s led to an Arab rebellion against the British that lasted for about three years. The great increase in the Jewish population became a source of real concern to the leaders of Palestine. This caused the British Government to issue an official policy statement in 1939 declaring that Jewish immigration and land purchases would be severely curtailed for five years and after that time it would be banned. With the possibility of a Great War in Europe looming so strong the British did not want to fight this war without the support of the Arab countries. This reversal of policy by the British was to gain Arab cooperation and seemed to guarantee an Arab rather than a Jewish state. We were in Palestine during the year of 1942 and while Great Britain never entirely stopped Jewish immigration, they did hold down quotas significantly. This was a move to appease the Arabs, which seemed effective at the time. Remembering the press interview in 1938 and the statement that I made, "The youth of today feel this War is Europe's problem." They need to solve these problems by themselves, but the solutions have not come. Perhaps, there are no easy solutions; otherwise these problems would have been resolved long ago. The attitudes of the two races are so distinct and different. The Jews are hard workers and want to make the land more productive. In contrast, the Arabs are content to sit on the land and let it remain as it always has been. This, of course, is an over- simplification of their problems. I feel that we should help those who help themselves. The US has played both sides on this issue, but the American people have, for the most part, helped the Jews to build a homeland for themselves.

The fact remains that this land has lived with constant conflict. We were fighting yet another war, with our planes based in their homeland in order to bring war to *their* enemy. True we were here by invitation to protect their land from being overrun by the Germans and Italians.

The people, both the Arabs and Hebrews, were very warm and friendly to us. I would say we were well received. We were much impressed with the new Jewish residents and their desire to work, and to work hard, to improve the land. We saw plenty of evidence of this in the tilling of the soil, the removal of stones from the fields and the building of new homes, hospitals, and businesses. The Arabs seemed mostly content to raise their animals, goats, sheep, donkeys, and camels and to leave things as they had always been. They enjoyed sitting around and having their coffee, a big social event. At the same time we saw the Jewish farmers raise great vegetables and fruits, which were plentiful and delicious. There was an abundance of grapes, figs, dates, and olives. I had never seen such large clusters of grapes and they were very good. The oranges were the best and sweetest I had ever eaten. They peeled easily and were sweet and juicy. We loved sending a bag of oranges on the planes when they went out on a mission. At high altitudes the temperature is so cold the oranges came back nearly frozen. What a treat! We really enjoyed the food while we were in Palestine.

After about two and half-hours of driving, we arrived at the great city of Jerusalem. When we left the coast we drove the last 50 miles over rolling hills, which reminded me of our own Ozark Mountains. There weren't as many trees as we had in the Ozarks but it made me homesick anyway!

When we arrived in Jerusalem, I felt as if I was in the presence of God. I remembered Psalm 122, Vs.1 that said, "Let us go to the house of the Lord." Just to be there and see the places and buildings where the events of history made such a great impact on the world was like being a part of the scene. I could hardly wait to take it all in. The new city of Jerusalem was much like most modern cities with fine restaurants; shops and good looking buildings but Old Jerusalem still had the ancient wall around it and appeared unchanged from ancient times. I set out with my friends to find a guide to spend the day with us and show us all the historical sites.

We were really fortunate to find a young man (about 30 years old) who spoke good English and was a member of the Franciscan Monks who were caretakers of all religious sights in Palestine for the Catholic Church. Here was a man so knowledgeable of all of the places and events that it was a real pleasure to hear his descriptions and significant details about what we saw. Our guide, Fr. Andrew, was wise beyond his years and gentle in his spirit. As a monk he had retreated from the world for religious reasons and lived under strict vows of poverty, obedience, and chastity. His presence added to the authenticity of the event. He spoke with such authority we knew what he was telling us had to be true.

This was a perfect time to have a tour of the Holy Land, as there were no tourists. Other than the natives, we had the city all to ourselves with the exception of a few soldiers who were around. Fr. Andrew began by taking us to Bethlehem to see the birthplace of Jesus and then guided us to all the sacred sites around Jerusalem. My first impression was how close everything was. In the Bible when it talked about "a journey into Bethlehem," I had visions of a long trip. Not so, Bethlehem was only three miles from Jerusalem. It seemed everything was close by. For example the Garden of Gethsemane was on the Mt. of Olives and this was just across the Kidron Valley from the walls of old Jerusalem. When we arrived at one of the sacred sites we would find a building which they called a Church to mark the location. These "Churches" were unlike our concept of Churches in that there were no seats or pews or congregations. For me, one of the most amazing sights was the altar that had been erected where the crucifixion occurred. There on display in locked cases were an enormous amount of personal wealth in the form of watches, jewelry,

diamonds, other precious stones and bracelets that had been left there by visitors who had been so moved to do so. This made a deep impression on me. It was hard for me to believe people would make such great sacrifices. I know their spirits must have been significantly touched by what they had seen in order to give up their precious possessions.

At the end of the day, Fr. Andrew took us back to the Dome of the Rock for a group picture to remember the day. I thanked him for the time he spent with us and told him how good it was to have such expert information about what we had seen. We developed a good rapport with him during the day and had exchanged many thoughts and ideas.

Fr. Andrew said, "It's good to be with Americans once again and hear from home." My startled look must have revealed my surprise to hear these words from him.

"Wait a minute, Father. Are you trying to tell me you are from the United States?" I asked.

"Yes, I grew up in Kansas City," he answered.

Mumbling to myself I commented, "That is the last thing I expected." Sensing my disappointment he began to explain, "I was a member of the Catholic Church and an altar boy. When I finished High School I was struggling with an identity problem. I was not sure who I was or what I should be doing. I enrolled in Kansas State and still could not find the answers I was seeking. I struggled with the secular world and religion before dropping out of school. I didn't have any peace with myself until I decided to become a Monk. Finally, I knew what I had to do. I took the vow of poverty, obedience, and chastity and lived in self denial."

I said softly "You had to give up your family, your freedom, and your ability to earn a living? Wow, what a sacrifice."

"No Sergeant. You see, when you are doing something you want to do, it is not a sacrifice," he replied.

"This may be true," I said, "but I can't

help but think about what all you have given up in order to do this."

"I am afraid you have missed the point, Sergeant. It is not I that is important." He continued, "It is what you have seen today that may strengthen your faith in the Lord you serve. That's what is important,"

This statement humbled me, and yet I felt encouraged by his spirit. We said goodbye to Fr. Andrew, and went off with my friends from the Squadron to the YMCA in Jerusalem where we made arrangements to spend the night. I was surprised to find a "Y" here, but it was almost new and magnificent. We were told it was built and equipped by some wealthy Americans. I was very thankful that someone had the money and was generous enough to build this splendid structure here. It was far better than anything we expected.

That night, I thought much about my parting conversation with the man with whom I spent most of the day, Father Andrew. I didn't know what the future would hold for me, but somehow I felt this experience better prepared me to face whatever might come in the future. I know the experience of that day made me a stronger person. I understood why Fr. Andrew was serving the Lord in the Holy Land. He would touch other lives as he had touched mine.

Soon after returning from my trip to Jerusalem, one of the airmen came to me with a problem. Because we left the States in such a hurry, many of the soldiers brought with them some very heavy burdens. Other men were dumped upon by their loved ones back home after their arrival here in Palestine. There were many personal adjustments that had to be made. They not only had to cope with the hazards of war, but also their personal problems as well. It seemed everyone had some kind of a problem and they needed someone to turn to. For them it was a difficult time of adjustment.

The unofficial Chaplain of the Group was not a position that I chose; rather it was thrust upon me. I suddenly found myself with a number of their problems. These young men (just boys) needed someone to listen to their concerns. Since I was two or three years older than most of them and had always been a good listener, I could relate to them. As time went on, I found myself punching more and more "Tough Luck Cards."

One of the disturbed airmen, Ted, came to me one night and jokingly said, "Sergeant, I sure would like for you to punch my T.L. Card tonight. I have had a rough week."

I said, "Sure Ted, tell me about it."

"Well," he started, "I got this letter from my wife and, well I don't know, it is kinda hard to talk about it."

"Okay," I said, "If you had rather not talk about it, we'll talk about something else."

Ted answered, "No, I need to get it off my mind and talk about it, but I don't know how to start."

I said, "Come on let's go take a walk, maybe the outside air and the bright stars will clear things up for you."

We walked silently for awhile. Then, Ted began to tell me what was bothering him. Slowly he said, "We have been gone from the States for about six months now, and the other day I got this letter from my wife saying she is three months pregnant. The dates don't all add up right, and I know this is not my baby. She said it was an accident and would give me a divorce."

"Do you want a divorce, Ted?" I asked.

"I am so filled with anger and hate that I don't know what I want," he cried. "I just wish I was back home and I would straighten things out."

"Being back there, Ted, is not one of your options, at least not just now. We need to talk about what options you do have and what you can do about it now," I told him. "Your wife offered you a divorce. For you this is one way out. Is this what you want?"

Ted was a long time in answering. "No, not really," he said.

"Is that what she wants?" I asked.

"I don't know what she wants." Then he added, "In her last letter to me, she offered to have an abortion, if that was what I wanted."

I asked, "Does this seem to be a fair option?"

"Not really," was his answer.

"Does this solve anything, Ted?" I asked

"Not for me, it doesn't," replied Ted.

"Tell me a bit about what kind of relationship you had with your wife. How long have you two been married? How long had you known each other before the marriage?" I asked.

Ted answered, "We have been married a little over a year now. We went to school together and have known each other for a long time."

"Well it seems to me that you really love her, in spite of her mistake."

"Oh yes, I do," he answered quickly.

I said to him, "Then you have just provided your own answer, Ted. If you really love her, you two can work things out."

"Love is the basis for working out your problem. I have not seen many problems that could not be solved by love. Tell your wife that you love her that you forgive her, and that you do not want a divorce. Tell her to have the baby and you will learn to love it as your own."

This is what Ted had been saying to me only he could not see it. This young soldier was hurting so deeply, that I stayed in close touch with him for a while. He and his wife eventually got their problems worked out. As an airman with one of our Combat Bomber crews he flew his fifty missions okay and was ready to be rotated back to the States. Before he left he came by to tell me goodbye and to thank me for helping him through a crisis in his life. I pointed out to him that he had provided his own solution. I wished the best for him and his wife, and told him that I would keep them in my prayers. Later when he was back in the States, he and his wife both wrote me a letter, thanking me for helping them out in a difficult time in their life and included a picture of their baby. What a good feeling it was to know someone appreciated

my being there when they had a personal crisis. They went through some troubled waters and some very difficult times, but Ted and his wife discovered just how much they meant to each other and how much love was there. Before Ted left to go back home, he was doing all right and I didn't have to punch his "Tough Luck Card" any more. In time, Ted had adjusted well to a stressful situation.

These were tough times for many of us; adjusting to a new life, in strange circumstances, in a foreign country and in an army environment. To add to this problem, many of the GI's were just kids away from home for the first time in their life. The Air Corps, which was part of the Army at that time, had to make men out of these boys in a hurry. No wonder so many of them wanted to come by and discuss their problems and have me to punch their "Tough Luck Card so to speak. What they really needed was some one to listen and to understand their problem.

Our combat crews left out early in the mornings to go on the bombing missions. After an early breakfast they went to a briefing that gave the details of the day's mission, then were taken out on the field to get into their plane for their day's work. Each combat crew had a plane assigned to them which they were permitted to name and keep for all of their missions. Slipping into their plane at the beginning of the morning the flight crews always found everything in readiness. All of the guns were loaded, cleaned and checked out, the bombs were secure in their racks, cameras loaded and serviced, fuel tanks capped off, and all of the trash from the previous mission had been cleaned out. The combat crews always had a good feeling about this. What brought about this metamorphous of change? An indispensable team which, day after day, was helping to carry the air war to the heart of the enemy.

As soon as we reached 20 thousand feet or so on our way to the target area, we occasionally had vapor trails to form. They would flow off the trailing edge of the plane's wing. This was a visible trail of water and ice crystals formed when the exhaust of the engines injected water vapors into the super cold air. When leaving the planes this vapor started at a fine point and gradually spread leaving a long defused trail behind. From the plane it was a beautiful sight to behold. I thought of "Happy Trails to You," the song Roy Rogers and Dale Evans sang as their theme song, but these were not "Happy Trails" for us. From the ground these white streaks across the blue sky stood out like marks that said, "Here we are." To see this vapor form was an uncomfortable feeling for us, but there was nothing we could do about these identifying trails.

When the combat crews returned with their planes some ten to fourteen hours after take off, they headed for a debriefing, a quick meal and off to bed, exhausted. This is the time the ground support team went to work. The Photo Lab processed the day's film, identified and printed them, reloaded the cameras, and took them back out to the planes sometime after midnight. There

they found the mechanics going over and repairing anything needed, the ordnance men bringing out the bomb load, and a crew from armament installing them in the plane, going over all ten guns of the plane, making sure they were in A-1 working order. Sometimes these guns had to be repaired before they went on another mission. Often the bombs had to be changed out at the last minute because of the weather over the target. Operations might shift targets and needed a different kind of bomb for the new target. This procedure took all night, but when the ground crews finished and the plane was ready to roll the next morning, the men from armament stood guard over the plane until the combat crews came to take over. The ground crew's work was over until the planes returned and it started all over again. While the planes were in the air, the ground crews tried to take care of some of their personal needs plus get some rest, unless there was something left over from the previous mission. That took precedence over every thing else.

*Combat crews leave briefing room with instruction on day's mission.*

# CHAPTER FOUR

### Egypt: Ancient Mysteries And Modern Ways

Just a couple of days after returning to base from our Jerusalem trip, we had our first air raid. In the early hours of the morning, before daylight, enemy planes flew over our airfield dropping bombs. They missed the airfield hitting only some orange trees at the end of our runways. It didn't even excite me enough to get me out of bed. For this attitude I was grounded by the Officer of The Day. For ten days I could not leave the base. The Lt. said I set a bad example for my men and I should show more concern. Big deal! I had not planned to leave the base anyway.

As soon as the ten-day's ban was lifted, I requested an overnight pass and went into Haifa for some good food and rest. Besides, I needed to talk to Sgt. Willem and his fiance alone about their marriage plans. His request had been submitted a long time now, and I felt it was getting close to time for the approval to come through. I needed one more chance to try to get them to see the futility of their getting married during the war.

Back in the States when I was transferred from the 154th Reconnaissance Group to the 98th Bomb Group, I had a full month's salary coming to me and I signed papers with the squadron to give my pay to Sgt. Bill Roseberry. I knew I could trust my cousin to get the money to me, since I didn't know where I would be stationed at the time this money came in. Bill received my money and sent it on to me in Palestine. As soon as I received it I decided I would blow the whole month's wages on Sgt. Willem and his fiance, Marsha, for one last chance to try to talk some sense into them.

Marsha picked both of us up and took us into Haifa. There I checked into a Hotel and invited them to come back and have dinner with me. Later they joined me for a special dinner served on a white tablecloth, linen napkins and fine silverware. While dining we enjoyed an excellent Russian orchestra playing, enjoyable and romantic music. Marsha was a beautiful girl and I could understand why Herman had become so infatuated with her. Marsha was an intelligent girl and spoke very good English. With her conversation, she added much to the enjoyment of the evening. This was the first time in six months that I felt like I was still part of the human race. What a great feeling to be there and be able to enjoy the surroundings, the food, and the music with good friends. I wanted to discourage any thoughts they had of marriage at this time since Herman was still in service. I tried to get them to see the wisdom of delaying their wedding plans until he returned to the States where he could take care of her in a proper manner. I found they were two very determined people, and wouldn't listen to a word I said.

While alone with him, I asked the "What if" question. I asked, "Herman, just suppose that after you and Marsha get married, you send your wife back to the States. What if it took two or three years for you to join her, and during that period of time your wife fell in love with another man while waiting for you. What do you think would happen?"

I could see how uncomfortable he was while I was talking to him.

"I don't want to think about that," he replied and then added, "I would appreciate it if you did not bring this up again."

I struck out on all accounts. He would

not listen to me, so I decided I would not interfere with their plans. The next day we learned we were to move out and leave Palestine within a week. Goodbye, Palestine and Hello, Egypt. Herman was frantic about leaving Marsha, but the French had a saying for this, and other situations: "C'est le guerre." So goes the war!

We moved about 100 miles southwest of Cairo to Kabrit, Egypt, which was an established RAF air base with runways and permanent buildings. At the time we moved there it had been the most bombed airfield in the Middle East. For protection sandbags had been stacked to a height of ten feet around every building and every spot a plane was parked. The airfield was kept well repaired and was in good condition. We were pleased to be there and had a perfect Lab set up for processing our films. It was a large building with living quarters for the men working there. Our section had grown now to 15 men plus lots of American cameras and other equipment. I was especially glad to get a 4x5 Speed Graphic for taking high quality still photos. Still we did not have our trailer photo lab, but our American aerial cameras began arriving. At this point we had 22 large aerial cameras made in the USA and only 8 out of many more British cameras. There were four squadrons of planes for whom we were responsible for processing films from their bombing missions. We had to work two shifts, one day and one night shift to give the coverage that was needed. What a far cry this was from our meager beginnings. We moved to this base on November 10, 1942, and remained there for three and a half months giving strike support to the British. Successful raids were made on Tripoli, Naples, and Tunis where a heavy cruiser and several other ships were sunk in these harbors.

During this time I managed to go into Cairo, one of the largest cities in the world, and also very intriguing. On my way in along the Suez Canal I observed men walking with long ropes attached to headbands, the other end attached to a ship they were pulling through the canal. The men leaned in sharply to the harness and it looked like they were under great stress as they pulled the boat along. I was impressed at how primitive this was as well as the other agricultural practices I observed. They continued to do things the same way they must have been doing them for hundreds of years.

We loved Cairo. This was an exciting city with great wealth and great poverty. Modern buildings and old slums; full of intrigue and mystique enhanced by night and the noise. Because of the war and the possibility of air raids, all city lights, and even the taxis, had blue lights and drivers who kept their hands on their horns. The dim lights made it a little difficult to see to get around, but increased the mysticism and excitement of the city.

To the parched and weary desert fighters of the Allied Armies, a furlough to the Egyptian capital of Cairo meant more than a brief escape from combat. It was that and more. Cairo was a rare oasis of luxuries from the exotic to the simple pleasure of taking a bath. It was easy to forget the war for the moment as we roamed the noisy streets teaming with people and cars and searching for bargains amid the Arabic cries of vendors peddling their wares.

One of the great hotels in Cairo at that time was The Shepheard Hotel, which was run by Swiss investors. The visiting press liked to stay here and they came from all over the world to cover the war story. The hotel was very beautiful with excellent accommodations. There were large rooms, the food was very good, and the services were excellent. Stationed throughout the dining areas and the bar were black men in native costume with a large fan on a pole that kept the air moving. This not only helped in the circulation of air, but helped to keep the flyies moving around. This added a very unique touch to the atmosphere of this setting. Their bar was one of Cairo's most popular watering holes. It was filled with British Officers, Free French, and civilians, some of which were obviously Axis spies. Since the news from the desert front was slow in coming to Cairo, in its absence, rumors prevailed. Axis spies haunted Cairo seeking information that might help Rommel on his drive toward the city. The City of Cairo had escaped the damages of war. About the only evidence of a war were the soldiers that freely moved in and out of the city, and the ambulances that brought new casualties from the bitter battles being fought in the desert to the hospitals located here.

As an official Air Force Photographer I had press credentials which permitted me to stay at the Shepheard Hotel, if I chose. On my first trip there, I had a

*Attack on Tripoli.*

35mm camera stolen which I brought with me from the States. I checked in the Hotel and went to my room, which was very large and contained a huge wardrobe for my clothes. After washing up, I hung my clothes in the wardrobe together with my camera. I felt safe because there was a lock on the wardrobe cabinet. I locked the wardrobe closet, took the key with me and then locked my room door. I went down stairs for coffee and was gone for no more than 15 or 20 minutes. When I returned the camera was missing. Nothing else in the room had been touched. The room and wardrobe were both still locked. Slick as a whistle, the camera was gone! Whoever took the camera knew what they were doing, and knew what they wanted. I don't think it was the camera they wanted as much as the film inside. I'm sure they were disappointed when they processed the film and found only "tourist photos," and nothing to help a poor spy looking for information. After that experience, I didn't trust anybody around the hotel. I felt like I had to be on guard all the time.

Somewhat daunted I went out shopping, or rather looking because I didn't have the money to buy another expensive camera. I was in a plush upscale jewelry store when the owner came up and started talking to me. He asked lots of questions but I was careful not to give very clear answers. I felt that I had to be very careful in these situations because I never knew who wanted to know, and how the information would be used. The first thing they always asked was, "What outfit are you in?" and then they asked, "When are the Americans going to bomb Ploesti?" I always acted dumb and asked, "Where is Ploesti?" If they answered, "Romania," I would say, "Never. We don't have planes that will fly that far." If they didn't know, I said I didn't know either, and that ended that conversation.

After a time he said he really had a good proposition to offer me. Still thinking cameras, I said, "What do you have?"

He said, "Do you see that beautiful lady over there? She is my sister and we want to get her to the United States. If you will marry her and send her back to the States we will take care of all expenses. Then when you get out of service you will never have to work again for we will see that you will always have plenty of money. Will you do this one thing for us?"

She was young and truly a beautiful woman. She could have easily been a professional model. I thought, "Why would I do this for them? They didn't know me. They must have a reason why they were so anxious to get this girl to the States." This offer, as wild as it was, never really tempted me. My dad always told me, "Son if anything sounds too good to be true, it probably is."

"You have a remarkable plan but you have the wrong man. Sorry," I replied.

Not wishing to discuss this any further, I left his shop. I couldn't help but wonder what he had on his mind. So many people live here from so many different countries. A large number of French, Germans and Italians lived in Cairo. Security in this city was very relaxed. With all of the mysticism, and intrigue, I couldn't help but wonder what was their ulterior motive. Could it have been for espionage? Perhaps. I didn't stay around to find out, and I will never know.

I thought about Sgt. Willem and my advice to him about trying to marry someone overseas. The long delay and not knowing when or even if you will return to the States is not worth the stress. Oh well, anyway, I was not ready to receive a proposal for marriage from the "brother of the bride."

I returned to the Shepheard Hotel to freshen up before going out to dinner that night with some friends I had met at the Hotel. In addition to the many reporters representing various newspapers from all over the world there were large numbers of military personnel, so it was not difficult to pick up all kinds of stories and rumors. Some of them would be true and some of them were pretty wild. I would just have to figure out which was which. One of the most plausible of the wild rumors around the hotel was that Adolf Hitler's security staff had reserved two floors at this hotel for the surrendering of North Africa and Egypt to the Axis. Hitler would be there in person to sign the acceptance papers. The way the War had been going for him, I am sure he thought this would happen and it would happen soon. As egotistical as Hitler was, this sounded plausible to me. So, I checked this out with a member of the hotel staff and was told Hitler did indeed have two whole floors reserved there at the Shepheard Hotel. The dates were left open, and they could not tell me the purpose of the event.

Egypt had broken off diplomatic relations with Germany, but had not declared war. Among the native population was a small, but not effective, group of Axis sympathizers who were pulling for Rommel. The students were most vocal and wanted to see their country rid of the British Army. They shouted chants of "Press on, Rommel, Press on! They were waiting for an invasion of the City of Cairo - an invasion that was never to come. While they waited, Cairo remained a great city of mystery and intrigue.

There were more and more GI's coming to Cairo for sight seeing, and R & R. The young native boys around 10 to 14 years of age worked up a pretty good racket for them to earn some money from the GI's. They would stroll the streets in groups of three or four boys shouting, "Shine shoes GI, buck-shees." In their language this meant "something for nothing" or for free. They would worry you to death yelling, "buck-shees, GI, buck-shees!" But believe me you had better pay them something for if you did not, they would flip polish all over your clothes and take off running. You could never catch them, as they knew the streets too well to be caught. True, there is nothing free. It was best to never let them shine your shoes, but go ahead and give them some money to leave you alone. What a racket these kids had going and there wasn't much the GI's could do about it.

The theater we went to that night was something to behold. I was astonished to see such a modern theater here in Egypt. I had never seen anything like it anywhere. They were showing American movies with Egyptian sub-titles. Like most stage productions, the movie was broken with two intermissions of thirty minutes each. The night I was there they were showing one of Cary Grant's films. We were seated at tables with small blue lights that came on at intermission. At that same time, the roof of the building rolled back in sections, telescoping into each other. It was a startling sight to see this happen and then sit there, under the stars, having refreshments. Looking at the big beautiful black sky, the atmosphere was so clear and the stars seemed especially bright and very close. At the end of intermission, the roof rolled back into place and the film continued. What a wonderful way to view a movie.

Within a few days after this, Sgt. Willem received approval to marry the Jewish girl he fell in love with in Palestine. He caught a ride back to Palestine so they could be married. Our Commanding Officer granted him one week to report back to our base in Kabrit. Willem made it back okay as a married man, saddened that it was so short a stay, but happy at last to be married to Marsha. His next project was to get her back to the United States to his people in New Orleans.

Field Commander Montgomery began to build up strength and prepare for a major attack on the enemy at El Alamein. The battle line, only about forty-five miles long, was drawn. The British had stopped the onrush of Gen. Rommel and his German tanks there. The battle line was not long for the Mediterranean Sea was at the north end and the desert with its deep sand was at the south end. Here began one of the most clever and deceptive battle plans that I have ever known. By day all was quiet on the front, but at night all was very different. Gen. Montgomery had only a few tanks on the desert end of the front line but over loud speakers,

31

he played recordings of large tank movements. These recordings were played at night while the few tanks he had ran about making as many tracks in the sand as possible. To further carry out this deception to the south, he established two dummy field regiments complete with dummy heavy artillery, which were made of lumber and covered with canvas. To complete the deception, the British also made dummy latrines and a dummy water pipeline made of empty gasoline cans. The Germans would fly over in the mornings to take pictures and would see all of the tank tracks and artillery build up. Because of the mass troop movements they had seen and photographed, they assumed this was where Montgomery would strike. They watched this happening and finally took the bait. The German General Rommel was known as the "Desert Fox" because of his quick and cunning ways. He had defeated the British all across the desert time after time with his quick and decisive actions. But this time, the old Desert Fox was out foxed himself. Rommel, thinking that Montgomery was building up his forces to the south at the desert end to make an end run around the battle line, began to move his tanks around to cover this movement. Montgomery continued to keep up this ruse until most of the German tanks were concentrated at that end.

From August to October the British built up reinforcements at the north El Alamein front line. The coming battle was to eclipse any battle the desert had seen before. The Allies had over 40,000 men there, mostly British Army, 800 artillery guns, and more than a thousand tanks, 300 of which were the new 36-ton Sherman tanks that Roosevelt had promised to Churchill. The tank commanders pinned their hopes on these Sherman tanks from the USA because with their 75mm guns they could out shoot anything the Germans had in the desert. After the Germans began to pull their tanks around to cover the build up they had been watching, the British began their offensive drive along the Mediterranean Sea coast. It was the night of October 23rd when the British began the largest offensive strike of the desert warfare. A thousand-gun artillery barrage that lasted for five hours, blasting through the elaborate mine fields the Germans had set up broke the silence of the desert night. The British soon broke through the line with the Axis out numbered and out manned nearly 2 to 1. The Allied forces completely surrounded the German tanks and had the Axis forces bottled up. Like shooting ducks on the water, the British completely annihilated the German tanks and Rommel's forces had to retreat. Hitler had asked Rommel to stand firm; and he did for a brief time. After only twelve days of fighting, the Axis lost most of their 500 tanks. Rommel was embittered by Hitler's "victory or death" order to stand firm. He said his troops stayed there fighting twenty-four hours too long, for it was then most of his men and tanks were lost. The Germans began their retreat on November 2, and by November 6th the British had driven them from Egypt's soil. By British estimates this offensive battle left 10,000 Axis soldiers dead, 15,000 wounded and their fleeing army left behind 30,000 men from the Afrika Korps, most of whom were Italian troops Rommel abandoned. This was the beginning of the end for the once mighty Afrika Korps. The war and history had made one of life's momentous decisions.

While we were still at Kabrit, Egypt and the war was looking better for the Allies, some American reporters arrived who wanted to get first hand experience and information. Our base was an excellent place and convenient for the reporters to obtain this information. They could headquarter at the Shepheard's Hotel in Cairo and come out to our squadron for reports. Most of the pilots who were flying missions were young, around 22 to 25 years old and they loved to kid around and have fun. When members of the press showed up for information or to fly a mission with them, often the reporters would be frightened off. Once when they were gathered at the briefing room, just before the mission was to takeoff, the crews started talking about the pilot's being so blind he couldn't find the target on their last mission. The pilot spoke up and said, "Oh, hell it wasn't me, it was my bombardier."

The Bombardier spoke up and said, "Oh I could see the target as plain as that elephant on your flight jacket". The flight jacket had an eagle on it. Needless to say the Press Corps would either have another assignment or they became ill. Suddenly they were unable to go on the mission. Their excuse was they had obtained all the information they needed. They would then take off back to Cairo and the safety of the Shepheard Hotel.

We followed close behind Montgomery and moved into the desert at a god-forsaken place called Gambat. The first night we were there I slept on the ground and covered myself with a tarp because the desert nights were very cold. The next morning I woke up with water on the tarp. Water in the desert? I couldn't figure that one out until I decided condensation must have formed during the night.

In April we moved to Lete, Libya, where paratroopers from the sky attacked us. A Luftwaffe Bomber flew over at a very high altitude dropping out men. It looked like ten paratroopers bailed out over the airfield where we were stationed. The attack came right at noon when we were standing in line to go into the mess tent for lunch. We saw the plane but the men didn't open their chutes for a long time, and it was hard to count how many parachutes

*All that was left of a nearby tent when German paratroopers blew up a B-24.*

*Searching the desert for German paratroopers.*

there were. With the long free fall, it was not easy to see the men until their parachutes opened. When this happened, we knew we had trouble.

The CO called for all men to come out to help round up and capture the German paratroopers. We were instructed to line up at arm's length apart in a single file and walk across the desert where the paratroopers had landed. That afternoon the whole squadron kept sweeping out across the desert until we had rounded up six of the Germans. We were not really sure how many made it down. The sun was going down and it was now too dark to see so we gave up the search. Since we were not sure if there were more men out there or not, we doubled security for the night. Guards with rifles were scattered all through the camp. I would hear a guard challenge someone: "Halt! Who goes there?" We had a special password issued for the night, and you better give the right answer. It was a very dark night with no moon. I had work to do at the photo lab to get ready for the next day's mission, and it seemed that I could not move around camp without being challenged,

Still, even with this tight security, the Germans were able to get to our planes to do their dirty work. They were skilled all right because they knew just exactly how to get around our tight security.

They also knew how to sabotage our planes and totally destroy them with very little effort. And they did it so simply. They had small incendiary bombs about the size of a fountain pen and they knew exactly where to place these little bombs so as to rupture the fuel tanks and set them afire. After midnight they destroyed four of our B-24 planes. What a terrible sight for us as we watched our planes go up in flames, and could do nothing to stop it. Our fire trucks were busy fighting the fires, but they did little good. The next morning with four of our B-24's still burning and smoldering our field was a sad sight to see.

Although the German paratroopers did considerable damage to our planes that night, they paid a very high price for what they had done. Sometime later that night, or early in the morning, the same small cadre of British trained Zulu soldiers we picked up in Durban, SA, pulled their knives from their scabbards and went out looking for the saboteurs. They found four of the soldiers before we did and castrated them with their knives, leaving them still bleeding, lying in the desert. They had drawn blood, so they were able to put their knives back in their scabbards and move on. Their job was done. When our troops arrived and found the condition of these men, we picked them up and transported them to a Field Hospital for treatment. After that the Military Police took charge of the German saboteurs. We felt thankful for the cadre of Zulu Soldiers we brought with us. They were there when we needed them.

We continued flying missions with our remaining planes hitting targets mostly in Sicily and Italy. On June 12, 1943, our planes flew a mission over an airfield in Sicily. One of our photographic crew members, "Doc" Hendricks, was on this mission. He was flying with the crew of Lt. Clarence Gooden 22, of Waycross, GA. As they were coming off their bombing run, Lt. Golden told us what happened: "The sky suddenly blackened with enemy fighters. One of them caught our No. 4 engine with a direct hit, setting it on fire. It was not long before No. 3 engine burst into flames. By now we were all alone. We didn't have enough power to stay with our formation and they went along without us. And let me tell you, it was a lonesome feeling with those 15 fighters on our tail coming at us from all directions with their guns spitting right into our faces. But we were not exactly idle; our gunners were in this fight too."

Everybody in Gooden's command was on the offensive against the attacking fighters. Lt. Gooden continued, "I made my co-pilot, Lt. Don Johnson of Akron, OH, and my radio operator, S.Sgt. Dan Kreutzer of Derby, CO, stay up front at their stations when the fighting first started. Bullets just missed my head twice, so I figured they might be needed. Later Kreutzer went back and manned a gun and joined the fight. Kreutzer, Cox, and Hendricks were wounded in the fight but refused to leave their guns. A 20mm enemy shell entered the plane and then exploded inside the plane wounding Hendricks. It did something to the rest of us to see those men fight back when it would have been so easy for them to give up. It made me proud I was one of them."

The crew gave those enemy planes a real fight. They fired 2,000 rounds at the fighter planes, shooting down five of the enemy planes and knocking one

of them "into the middle of next week." While all of this was going on Lt. Gooden was fighting the fight of his life to keep his plane in the air. Soon they only had one good engine left and the Lord only knows how many bullet holes in the ship. There were no brakes, no hydraulic system, no radio, and only one good tire.

"How did you make it to Malta?" Lt. Gooden was asked.

"Listen, brother, we just lead good clean lives, that's all."

Sgt. Hendricks was wounded in the head, neck and legs. Hendricks said when he no longer could man the waist gun because blood was flowing down into his eyes, he had to lay in the floor. While he was there, he gathered up the shells that had not fired and ejected from the guns. The waist guns would kick out shells every once in awhile that were not fired. He couldn't see them, but he could feel, and he started gathering them up in case they needed them. When we lost Sgt. Hendricks, we lost one of our very best men in the squadron . He was in the hospital for four months before he came back to us. His injuries were serious enough to have sent him back home, but he wanted to rejoin his old outfit. He was tough all right and fought hard for recovery so he could rejoin us.

The Island of Malta was but a flyspeck in the Mediterranean Sea, seventy miles off the coast of Sicily. It was a speck of limestone rock with three illequipped airfields maintained by the British, but had been a thorn in the side of the Axis. In desperation, Reich Marshal Herman Goring called for the total destruction of this Island from the air. Being so close to the airfields of Sicily this was an easy target for the Luftwaffe Bombers that were quickly dispatched to neutralize Malta, which became one of the most heavily bombed targets of the war. In 1942 bombs fell on Malta eight times a day and still the people of Malta and the British never gave up. By late fall of 1942 the Germans gave up on Malta, and shifted their planes to the Russian front. I was thankful that Malta never fell to the Axis, for it proved a

*Author stops to visit with nomads in the desert.*

life saver for many of our planes and crews.

We were still flying out of instant airfields that were nothing more than the sand scraped off by a Motor Grader, exposing a hard crust of dirt for our runways. The sand and dust were so hard on the engines, I don't know how they stood up to it. Of course each plane had its own ground crew of mechanics that kept them flying. When a plane returned shot up and damaged, they also made those repairs. When an engine was showing excessive wear, the mechanics would replace it with a rebuilt engine. At first we got all of these rebuilt engines from the States, but with so many B-24s in this area, the Army did something about this to speed up the exchange. A facility was put in at Cairo to rebuild all of these Pratt and Whitney engines. Of course the mechanics that did this work were all Americans, but a lot of Egyptians were used as helpers or laborers. The GI's started calling these rebuilt engines from Cairo, "Pratt and Wogs." "Wog" was our nickname for the Egyptians. These were very good engines and our flight line crews did a good job of keeping those B-24's flying.

While in the desert we ran across different tribes of people known as the Bedouins, or Nomads who are wanderers. In fact their name comes from a Greek word which means, "wandering around in search of pasture." These wandering tribes originally were from Arab descendants. They may stay in one place a few days or it may be for several weeks. When the food supply was exhausted, they moved to another source. They do not wander aimlessly for they know the territory in which they wander; where the water is and the types and kinds of plants there. There are two types of Nomads. The "hunters and gatherers" do not produce any food but must survive on what nature provides. These nomads acquire only enough of the necessities of life to survive and no more. Since they produce nothing, they cannot provide for an expanding population. The size of the tribe remains fairly constant. Then there are the "pastoral nomads" who are producers of food and depend on hunting as well as herding animals. Since they do produce food their tribal units do increase in numbers and varies accordingly.

With most of our film processing done at night we often became hungry and cooked up some good food. We soon found that the grounds we used for making coffee in our tent had a second life. After making our coffee, we would lay the grounds out the next day in the hot desert sun to dry out. When the coffee grounds were dry, we put them back in the package and sold them to the Nomads. Since they boiled their coffee and made it very thick the Nomads didn't know the difference. It was good coffee to them. No matter where you would be in the desert, if you stopped for five minutes, the Nomads would be there. Where they came from, you would never know. How they lived in the desert was a mystery to most of us. The Bedouins were nomadic herdsmen that wandered the desert with their animals and families. Appearing suddenly, they would ask for cigarettes and would try to sell you chicken eggs. Where they got the eggs was a puzzle to us, but they were good. We were glad they were there with their eggs.

Our equipment in the daytime would get hot enough to fry an egg in the sun, but we had a better way of cooking them. We had a rather ingenious stove in our tent for cooking at night. The "stove" was an empty five-gallon gas tank turned upside down so that the cooking surface was the bottom of the can. We used aviation gasoline for our fuel (what else was plentiful?). For safety we kept this fuel source outside our tent and ran a small copper tube under the sand to the "stove." Sometimes it was a little hard to regulate the flow of the fuel so we would not have too hot a fire, but it worked.

The camp cooks would always bake their bread at night and we would be processing the film at night from the mission that day, so we worked a deal with the cooks. They liked to take pic-

*Snow-capped Mount Etna, Italy, as seen by B-24 crew on way to target at Naples.*

tures, and we would process these in exchange for some hot bread. This really went good with the coffee and fresh eggs we picked up from the Bedouins.

We didn't get any American cigarette rations over there, only British. They came in small packs of five cigarettes and none of the GI's liked them. The men said that's why they came in such small packages, no one could stand more than five. A few of the men smoked them anyway, but most used them as barter material with the natives. They were good for this.

Along the edge of the desert both the British and Germans had built parts of a black top road to move their equipment. With the strong winds and shifting sand the road would often get covered so you couldn't tell where the road was. To get around this problem Arabs by the hundreds walked along sweeping with a broom constantly. They didn't have equipment to do this but native labor was plentiful. (Besides they didn't have any thing else to do.) Sometimes at night the wind would get ahead of them and deposit large quantities of sand over the road so the next day they would have to work hard even to find the road.

On one of our missions to Naples we had just about finished with our briefing. The weather report had told us what to expect over the target. The operations officer had briefed us on the target then asked if there were any questions. One of the bombardiers spoke up and asked if we had any problems over the target would it be okay to drop our bombs down the center of Mt. Vesuvius? There were two active volcanoes in Europe: Mt. Vesuvius, which was 4,190 feet high, and only seven miles from Naples. The other was Mt. Etna, nearly 11,000 feet high and on the tip of Sicily. We flew near both of them frequently. The mere thought of dropping bombs down these volcanoes sent Col. Kane into orbit.

"I hope this is someone's attempt at humor. I don't want to ever hear this mentioned again, even in jest. If you talk about it, someone will do it. Those volcanoes are strictly off limits even for a joke. The consequences would be too great for any one to even think about," Col. Kane responded. "Is that perfectly clear?"

There was dead silence in the briefing room. I think it was abundantly clear to every person in the room. It was never mentioned again.

# CHAPTER FIVE

## *North Africa: Perils and Victory*

We had no equipment for desert warfare. Again, we depended upon the British to supply us. They gave what they could, but at this point the war was moving at a hot and fast pace. Much of what we needed was picked up from the fast retreating Germans and Italians. The most precious item in the desert war was water. The lack of water posed a major challenge for everybody. We had no water tanks except what the Germans left behind, and that was not very many. Often, when we found a portable water tank we discovered that it had been shot full of holes when the Germans left. What little water we had was flown about 500 miles to us. With nothing to hold the water, we were rationed one canteen of water per man, every other day. This was for our total allowance, to be used any way we wanted to, including washing our body, shaving, drinking, or any thing else. I donated mine to the mess hall for coffee and would get back the equivalent in coffee. We often went two or three months without taking a bath and then we would fly back to Cairo to get equipment and experience the luxury of taking a bath in a whole tub of water!

On one of his trips back to Cairo on a three-day pass, Lt. David Browne, the officer in charge of our Photo Section, decided he would like to have a dog as a pet back at the squadron. Lt. Browne set out looking in the pet shops in Cairo for a puppy that appealed to him. He was successful in finding just the right one, which fit his idea of a real dog. He was an eight week old registered male Alsatian Hound, a particular breed of German shepherd dog that was an outstanding sheep dog. He was a fine, good-looking puppy. Lt. Browne paid for him and brought the puppy out in the desert where we were stationed, and named him "Alex." All of this was fine, except for one thing. Lt. Browne was a young kid from Detroit, Mich., who had never owned a dog in his life and didn't know the first thing about taking care of an animal. Since Alex was the only dog in the camp, he made an instant hit with all the other GI's, and received lots of attention from them. In fact everyone thought of Alex as "our dog," and he was a good addition to our outfit. We were glad to have him.

The Western Desert, where most of the fighting was done, was a rectangle 500 miles long and 150 miles wide. It was fringed by a fertile coastline, but the inner reaches were a vast wasteland. Besides the shortage of water, we had other problems in the great Libyan Desert. There were extreme heat and dust storms every day. It was not unusual to see the temperature get as high as 125 or 130 degrees in the day time with the wind blowing constantly in the

*The winds blew hard and lashed out at both men and equipment. The winds were loaded with cutting sand that stung.*

same direction; pounding faces so hard it would blister and sting. Sometimes it would be so bad I would hold my hand out at arm's length and could not see it. I even tried wearing my gas mask during one of these storms but my beard was too long and the mask irritated my face so much it was hard to keep on for very long. I did manage to sleep with it on for several nights because that was the only way I could get any rest.

At night, the wind would shift, and blow just as hard from a different direction. The temperature would drop down to the 50s or 60s; these were not unusual experiences. Temperatures fluctuated as much as 60 or 65 degrees in one day. Torments in the desert were many. Sand as fine as talcum powder routinely clogged our cameras and everything else. Driven by the hot southerly winds, the sand filled your nostrils and ears, so that you felt terrible. There was no water to wash it off. I don't even know why we were there, unless it was just a stopping-off point on our way somewhere else. The time spent there was inconceivable misery, but we were all in there together and there wasn't much complaining about our situation. I recalled hearing my mother say a thousand times, "What can't be cured must be endured." I guess that was our attitude. We sure couldn't cure our situation, so we toughed it out with much complaining. When things would really get rough, I diverted my thoughts about our situation by thinking about more pleasant times and good things that had happened. I remember thinking about the good foods Mom cooked and the hunting trips I would have in the woods back home with my Dad. This would take my mind off how really tough things were.

This barren, forsaken place called Gambat was forty miles southeast of Tobruk. The only sign of life was the scorpions. There were no village, no buildings, just endless miles of nothing and plenty of scorpions, desert rats, and flies. The scorpions had a poisonous sting that was quite harmful. They were not only plentiful, but they were big and

dangerous. It was almost impossible to avoid them, but I tried. One of their favorite places to collect was in boots after you had taken them off at night. I could almost certainly count on two or three in each boot the next morning. Their sting was so painful; I can assure you that I didn't forget to check my boots each morning before putting them on.

About the only thing worse than the scorpions were the real desert rats. They were also big, mean, and plentiful. I heard a couple of the fellows talking about the rats one day and one of them said, "I don't think I have ever seen a rabbit with a long tail before, but back in Texas where I come from, we call them varmints 'rabbits.' And they are so friendly they would just come right on into your tent and chew your leg off if you would let them." Some Texans have been known to exaggerate a little. The rats were neither as big nor as friendly as he would have you to believe. But they were mean and hungry. They were scavengers and the best way to deal with them was to shoot them on sight. Fortunately for us they didn't come around our tent very often because our dog Alex didn't like them and they didn't like Alex either.

Then there were the flies. Flies by the millions. Pesky, irritating, ubiquitous flies that got into everything. Even when you talked, you had to be careful they didn't fill your mouth or fly up your nose, there were so many. I remember how covered with flies our food supplies would be as they were being delivered to us out in the blue (That's what the British called the desert, "the blue.") I never did figure out why. Once I saw a rack of lamb, which was to be delivered to our cooks sitting on the floor of a truck so covered with flies it looked as though they were trying to carry it off. The Wogs kicked it out of the truck and it fell to the sand, and I thought that's what the Arabs think "dropping off" means. It was hard to have much of an appetite with your food handled like that. I knew though that was no different from the way they handled their own food for I had seen it in some of their markets. They didn't know what refrigeration was.

I don't know why or how this place got its name unless it was to designate an area. I don't even know why we were there unless it was just a stopping off point on our way somewhere else. The time there was inconceivable misery.

With the help of the British Army we established our camp at Gambat. We used their tents because they were very good for cover in desert warfare. The British tents were a light color, which blended in with the sand; were low to the ground with long sweeping wings; and contained a tent within a tent. This prevented them from getting as hot as our American tents, which were green and single layered canvas. Boy, did our country have a lot to learn about desert warfare! I understood from reporters that the US Army had American troops out in Arizona training for desert warfare. By the time they were trained, we hoped to be out of that hellhole.

Going through the battlefield at El Alamein, where General Montgomery started his long push across North Africa, we saw some startling things. It was here that the Axis Army stood, fought furiously, was defeated, and had to retreat.

I was very touched by the ugliness of war and the beauty of this battlefield. Let me try to explain this paradox. No picture, any painting could ever reveal more vividly this scene forever implanted in my memory. The shell torn, scattered earth was strewn with remnants of the tools of war like a lawn covered with the leaves of autumn. Trucks

*Equipment graveyard at El Alamein.*

*Graveyard at El Alamein.*

blown in every position, many beyond recognition of their former condition, battered tanks still standing staunch and erect, as if in tribute to their once great power, which now is lost. Guns standing as ghostlike guards in their empty emplacements awaiting crews whom would never return. They too have lost their great power to kill and stand as harmless as the dark shadows cast by their long barrels. There are trenches where men huddled to protect themselves and died; minefields to keep out the tanks; entanglements of barbed wire through which the wind sings a ghastly chant of war. Here was an airplane engine almost totally buried in the sand by its harsh impact with the earth, and over there is the wing that once sustained it in air, and perhaps beyond the fuselage once its support.

There are the other parts of the plane hardly distinguishable after the crash. Beside that truck, or tank, or gun, I saw a neatly arranged pile of rocks marked only by crude, but recognizable, crosses made from parts of the wreckage. Here lie ten thousand of those who have given their lives so others may share peace and happiness in a future yet to come. But the most amazing fact of all, here in the desert, nature tries her hand at concealing man's madness. Bright red, and blue, intermingled with varying shades of yellow are beautiful flowers full of life in a field of death. One wonder: Is God trying to hide our blunders, or is He offering a tribute to those who have died for the cause of peace? War goes on, and Man's inhumanity to man is there. Here stands El Alamein as a legacy of loss.

Our Photo Lab Trailer still had not come in and with water still rationed we were unable to process our film at this base. All film had to be sent back to Kabrit, Egypt for processing in the fine Lab we had just left. Did we ever miss that Lab! We had no lab work to do, and our only task was to try to keep the cameras clean and mounted on the planes for the missions. This was hard to do because the wind blew so hard that sand was in everything. It was difficult to keep the cameras operating. Our cooks also had a problem with the wind and sand. No matter what food the cooks at the mess hall served it was always sprinkled with sand. It was here that I started flying missions again as a photographer. I had flown on a half dozen or more missions when we first arrived in Palestine to show our crews how to work with the British cameras. Those missions were mostly strikes at ships and the harbors in North Africa. But flying missions now was different. This was the big time with all the anti-aircraft guns and pursuit planes over heavily defended targets. The combat crews were trained and worked together as a unit. I was not to replace anyone but to go along as an extra passenger; sometimes as a gunner, but always as a photographer. I asked for this duty. How else could I train gunners to be photographers too, if I had not been there and had this experience?

My first mission under these conditions was to Naples, Italy. Forty airplanes were sent out from our group in North Africa. We encountered some bad weather on our way and many of our planes had to turn back. When we were near Naples, only three planes had made it through the storms and bad weather. I was on one of the three planes.

I had never been so cold in my life. Of course the planes were not heated and at that high altitude the temperature was about 30 degrees below zero. We did have sheep skin coats, pants and boots, which helped, but still it was cold. At twelve thousand feet the air becomes too thin for breathing. There is not enough oxygen in the air, so we had to put on masks. By the time we got up to an altitude of twenty or twenty five thousand feet it was so cold that the breath exhaust valve on my oxygen mask kept freezing up. I would have to take it off and beat the mask over some object while I tried to clear it up.

We were about to make our bomb run over the Seaport of Naples, our target on this mission, when the pilot broke radio silence to communicate with his crew and the other planes. As he did this we heard the BBC loud and clear over our earphones. The Andrew Sisters were singing "Friendship," a very popular song back in the States at that time. The part they were singing as they came on the radio was:

"If you are ever in a jam, here I am. . . If you are ever up a tree, call on me. . . ."

That struck all of us as very funny. I wondered if that was some kind of an omen or a sign that we had a friend up there with us? How could that be, hearing the Andrew Sisters singing, "If you are ever in a jam, here I am" at twenty two thousand feet and at that particular time?

Then the pilot called on the intercom and said we were beginning our bomb run. I strapped myself to the plane, opened the bomb bay doors and got ready to take pictures of the bombs falling. I noticed the city began to come into view. It was a huge city; I guess the largest I had flown directly over. Down in the city it was twilight, but in the air we had full light. It looked so quiet and peaceful below with long and interesting shadows cast on the ground. But this thought soon faded as all hell broke loose in the air around us. Black puffs of smoke marked the spot where a shell exploded. Flak was all around us - each one meant for us. We could see the red flashes of the antiaircraft guns as they fired. Soon the black smoke became very thick, but we continued to fly through it. The Harbor below came into view and we began to drop our bombs. One of our three planes was hit over the target and I saw it go down. Then the plane I was on was hit in the bomb bay where I was taking pictures. The camera I was holding was hit by a piece of flak that knocked it out of operation with a direct hit. This was a big camera that covered most of my body with huge lenses and weighed about 35 or 40 pounds. It protected my body and saved me from a direct hit. I laid the damaged camera up on the flight deck and started walking across the cat walk back to the tail of the plane thinking, "We have it made now. We are off the target and out of that flak." I got to the back window just in time to see some of the German ME-109 pursuit planes waiting for us to come off the target. They lay back because of all the anti-aircraft flack over the target. The ME-109's were getting into position to knock us out of the sky. One plane was came down on us from five o'clock high as I saw tracer bullets from one of our guns going right into the engine of the fighter plane. Black smoke started to pour out of the plane. I was taking pictures of this with another camera when I stooped over to lay down my camera and pick up a machine gun to polish him off. The timing was just perfect. When I stood back up, there was a big hole where his 20 mm cannon had shot through the plane at the very spot where my head had been seconds ago. The shell was supposed to explode on contact, but for some unknown reason it went through the plane, making a hole big enough to put my leg through, and exploded outside the plane. This German pilot was so close to our plane, when we hit the engine of his plane, I could see him rip off his oxygen mask and deliberately tried to dive his plane into us. He missed and his damaged plane fell off into the sea. I lost sight of him and do not know if the pilot bailed out or not.

It was always German pursuit planes that came in on the attack. There were Italian attack planes up there with us, but they were always out of reach of our guns. I thought it was odd they never came close enough to get into the fight. I thought if I ever get to Italy, I

want to find out why they did not defend their own country.

It was night now and we had to fly back to our base alone. Our other plane was shot down by one of the ME-109's. Some thirteen hours later we landed back at base with a crippled plane and a very weary crew. When returning at night like this there was always the danger of missing the air base. It was very difficult to tell the difference between flying over water or the desert. If you missed the base, you just kept flying on into the blue until you ran out of fuel. The ground crews checked to see how much fuel was left in the tanks of our plane. There was not enough to measure. We must have come in on fumes, because there was not enough fuel to fly on. For security purposes our airfield was lighted at night with blue lights covered with cones. These cones had narrow slits cut in them so that they were only visible if the plane was at the right angle and the right height to approach the runway for a landing. That night we didn't have enough fuel to fly around and find the field. The runway appeared just at the right time to make this remarkable landing. As tired and exhausted as we were, we went directly into the debriefing room. This is where we had to tell all we knew saw, and felt about this mission. In the debriefing room, following the mission each of us was given only 1oz. of whisky, which was to help us relax. The whiskey was always kept under lock and that 1 oz. was all anybody could get. In the mornings before going out on a mission we were given a bowl of vanilla pudding and were told this would be good for our stomachs.

As we were turned loose to go to our quarters and rest, the officer in charge said to me, "Sergeant, turn around and fly again as soon as you can. This will be good therapy for you, and also will be of great benefit to you later on."

I replied, "Yeah, I guess it would," and went to my tent and fell, totally exhausted, into a deep sleep.

After a good night's sleep, some food, and rest, I felt much better, and man-

*Alex.*

aged to get in some fun time with our dog, Alex. He had grown into a good-looking dog and really did like to play with all the fellows. It was good to have a diversion like this dog because it took my mind off a lot of heavy stuff.

After a couple of days, I was ready to go again. This time it was with another crew and another target. Would you believe the ship was named "Arkansas Traveler" and the pilot was Lt. William Bacon from Fayetteville, AR? I kid you not. It was for real. This time the target was the harbor at Messina, Sicily, and was the closest I ever came to hanging my dog tags up forever. We took off and made the trip without incident until we got to the target. As we opened the bomb bay doors to start our run, we received a direct hit with flak in our number two engine. It caught on fire and we started dropping behind. The formation of our planes had to go on, as they could not fly as slow as we were forced to. I could see our co-pilot working furiously trying to put out the fire in the engine with the built in extinguisher. About twenty enemy pursuit planes saw us slow down with an engine on fire and thought we would be easy to knock off. So here came the Germans and their ME-109's just like chickens after a June bug. The first plane got high above and came down on us from the tail to the nose with guns blazing. He dotted our plane with a neat line of holes and knocked out our number three engine. By this time the co-pilot

*Nose art, Arkansas Traveller.*

had the fire out in number two, but we still had only two of the four engines running. The pilot broke radio silence and asked our radio operator to come down out of the top gun turret to radio our position to headquarters and to tell them we were going down. He told the other crew members to put on their parachutes and to stand by to bail out. I was alone in the bomb bay and heard this from my headphones, one of the few things still working. I walked to the rear of the plane and saw our gunner coming out of the turret in a panic trying to get his ëchute on. The Pilot had given the command to abandon ship so I stopped and began to help him. He was so nervous he couldn't get his chute on. After we got him all harnessed he thanked me and went over ready to jump. On his way out, I yelled to him, "Let's have spaghetti for supper tonight." He gave me a thumbs-up sign as he jumped.

After that a very strange thing happened. I had an out-of-body experience. It was an odd feeling facing death like

tation for anywhere I wanted to go. This was a good find for me and I enjoyed my bike very much.

With water still rationed, we didn't take a bath very often. Our clothes would begin to get stiff from the salt, oil and grime from our body. I wore them as long as possible and when I thought I couldn't stand it any longer I would wash them; first in sand to break up the crud and then rinse them out in gasoline, which was plentiful. With the sun and wind it didn't take long for them to dry out. One afternoon after doing his laundry, Lt. Browne decided it would be good to give Alex a bath. After all Alex had not had a bath since he was brought into the desert. Browne proceeded to wash the dog with the only thing he had available and that was 100-octane gasoline. Poor dog! It blistered the skin all over his body and his fine coat of hair all came off. I went to the medical section and got some ointment for burns to put on him. The problem we had out there in the desert was getting any kind of a sore to heal. If we had a cut it simply would not heal. Any cut or scratch we got in the desert stayed like an open wound. Try as hard as we could, we couldn't find anything that would help Alex's skin. The poor dog went from being a popular friend to something that every one tried to run off. He was a horrible, mange-infested sight. Lt. Browne came to me and said he was sorry about the dog but he couldn't do anything for him and would have to get rid of him. I asked the Lt. not to do that. I would take him and try to nurture him back to health. Lt. Browne then gave the dog to me and the challenge to help him was all mine. I continued to try to heal his skin but the sand would make a mess with the ointment I had rubbed on him. I put a blanket under my cot and he slept there. He looked so bad no one else wanted to even look at poor Alex.

The next mission for me to fly was back to Naples. We took off with a different crew early in the morning so we would be back by dark. Coming in at night it would be hard to tell whether

*Tech. Sgt. Blundell on an Italian motorcycle.*

find our airfield. We surely did not want to miss our base, which was so easy to do at night.

On our way up to Naples, we flew at a high altitude over land, so we took incendiary sticks that would burn through anything, and tossed them out of the plane ever so often. This way when we came back we would have fires to show us the way home.

On this particularly mission just as we were about to start our bombing run, all four engines went out and we began falling like a rock. We had not been hit. The engines all stopped at once. We started to throw out all the weight we could, even dumping our bombs. Both pilots worked feverishly trying to get the engines started. The pilot's oxygen mask froze up at the exhale port and he reached up and yanked it off. I knew he couldn't fly long at this altitude without oxygen, but since we believed we were going down I guess it didn't matter. The ME-109 fighter planes came over to polish us off. Black smoke poured out of each of the engines, so I guess they thought we were already done in and perhaps we were too low for them to attack but for some reason they tucked tail and flew off leaving us to crash into the sea.

When we were only about five or six hundred feet off the water, our pilots got two of the engines started. They pulled the plane out of the dive and continued working on the other two engines. We then leveled out and flew

48

*Capt. David Browne gives Alex a bath in the desert with aviation gasoline because there was no water.*

back across the Mediterranean Sea at that low level. We were never sure we would make it back without the engines going out again. We spent some anxious hours returning to base just skimming over the water and we gave some people on boats quite a shock, I'm sure. After about 13 hours from take off, we returned to our base, very tired and very weary. Believe me, it never felt so good to be back home.

Alex was still trying to recover from his sad ordeal of a bath in gasoline. When I went to the mess hall for a meal, he would follow me only to be yelled at by everyone, "Get out of here!" He looked so bad; no one could stand to look at him, especially at mealtime. Alex would tuck his tail and return to my tent. I always took some of my food back to feed the poor dog.

Ever since the time of Moses, the Land of Egypt has been plagued with swarms of locusts. When God told Moses to stretch out his arm and proclaim that the land of Egypt be covered by locusts, so it was then, and so it was now. We were no different. A strong east wind that blew all day and all night brought a cloud of locusts that covered the land. They were so thick they even covered each other. The sky was black with locusts. These migratory grasshoppers

travel in great swarms and feed ravenously on anything they can. About the only thing we could do while they were there was to stay in our tents for protection. We were issued netting to go over our cots to protect us at night as we slept. Even so, try as hard as we could, the next morning there would always be three or four of the locusts that somehow found a way to get in there with us. It was totally frustrating to us. Their presence brought a complete halt to everything we tried to do for a few days. Then we began to get philosophical about it all and decided that if Moses could put up with this, why shouldn't we. About that time they left as suddenly as they came. For this we were indeed thankful.

We had been through a very rough time and I'm sure someone up the line thought we needed some relief and decided to send us a treat. I don't know who came up with the idea but an airplane, not from our group, came in and landed with enough beer for the whole outfit. This was the first and only beer we had in all the time over there. The beer was unloaded and sat there all day in the hot sun. I guess they were trying to figure out how to ration it to us. At supper that night they gave each one of us a slip of paper good for one six-pack of beer. Of course it was very hot and we had no way of cooling the beer but most of the GI's drank it right then, hot as it was. Then some of the men, remembering the oranges from Palestine, offered to place the beer on the next mission for anyone who wanted to donate their beer. This was a riot because so many wanted to do this. Of course when the plane went up on a mission, all the beer froze and the bottles burst, making a big mess in the plane. This was one sad group of GI's who weren't able to taste the first beer they had seen in months. We didn't get any more beer over there. That was the end of that.

# CHAPTER SIX

## *PLOESTI: Planning & Performance*

We moved into the Bengasi area five months before our big mission on the oil fields of Ploesti, Rumania. Our planes were stationed at Lete, Libya, which meant we were now some two or three hundred miles closer to the Ploesti target. As important as this refinery was to the German Army, we couldn't help but wonder when we would get around to this mission. Of course nothing official came down to us, but we all knew Ploesti was out there and that one day, we would go. None of us in the 98th ever thought about the Ploesti mission being a low-level attack. Our planes had never been used for anything like that before. This low-level attack was designed to keep our planes from being detected by radar theoretically giving us the advantage of a surprise attack. We were told later that Gen. Dwight D. Eisenhower, the theater commander,

*B-24 lands in Bengasi on dirt runway after rain.*

*Bengasi - The rains came. Six inches of water in their tent. Cpl. Fisher, Sgt. Marcinko, Cpl. Whewell.*

approved the mission and the timing, but did not give his opinion on the low-level aspect of the attack. He wanted the job done, but was not sure whether the raid should be a high or a low-level attack.

More than six months in planning, this mission was assigned to Gen. Lewis Brereton's Ninth Air Force Bomber command. Two hundred B-24 Bomber planes were to make this attack with each bomb group assigned to take out a different part of the refinery. That way, we would be more certain of a complete wipe out of that installation. Col. John R. "Killer" Kane's 98th Bomb Group was designated to take out the main part of the refinery. What a huge responsibility for this group!

Look at us! We are the 98th Bomb Group of the 9th Air force who was selected for such an awesome task. Look around. Can you recognize us? We just came out of the Libyan Desert where we spent the last six months crossing it. There we were plagued with crawling scorpions, penetrating winds with cutting sands, and flying locusts that ate everything they could. Our clothes were tattered rags, a collection of American, British, even German and Italian uniforms left behind by a retreating army. Water was so scarce it was rationed. We had not had a bath or water to shave during six months in the desert. Our hair was long and raunchy and we looked and smelled like the "desert rats" that we called ourselves. Because the strong winds ripped and tore our tents, they were patched with scraps of anything we could find. Our shoes were coming apart; and held together with strings, tape or wire. We looked like the last remnants of a retreating army but we were not. We were the Pyramiders, the 98th Bomb Group of the 9th Air Force. We may not have looked like it, but we were still undefeated Americans and ready to take on the next challenge before us!

Having survived the worst the desert could throw at us, we began to focus on a more comfortable place to live. The GI's began to show the real stuff they

At Bengasi, there was sufficient fresh water and time to clean clothes and men.

After five months in the desert, a bath at Bengasi was a time of celebration. Cpl. Martin Potter assisted Sgt. Herbert Haggas in washing some sand off.

were made of, and how adaptable they still were. Tents were set up with floors made out of ammunition boxes. A shower was created with an elevated water tank. A bathtub was brought in from the nearby town of Bengasi, and one GI even made a washing machine out of junk he picked up. As you see, all of these improvised luxuries were related to water and to cleanliness. Our men had not forgotten how great it was to take a bath and have clean clothes to put on. Yes, we regained quickly what we lost in the desert, our pride in our appearance.

One of the qualities of life that brought us through the grimness and hardships of this war was our sense of humor. We never lost that. Struggling

*Sign erected at camp shows the good humor of the men. It was this sense of humor that helped get them through hard times.*

with the drudgery of the desert the men kept their spirits up and their attitude positive with their sense of humor and kidding around. When we got to Bengasi, some joker put up a signpost showing the way to different cities and other places. At the top of the post was an arrow that said, "Berlin 1,000 mi.," next came an arrow that said, "OSHKOSH," then pointing to a German airplane junkyard was an arrow to "Blunks Junk, 50 feet." Following these signs were some homesick cities: "Omaha, 6,500 mi.," "New York, 5,000 mi.," "Los Angeles City Limits, 10 miles." Of course we all got a kick out of this. It raised our level of optimism because we at last had a road sign to show us the way home. Besides it made us feel good to know that the Los Angeles City Limits were so spread out that it was only 10 miles away!

Now that we were out of the desert and in a better climate near the seacoast of Bengasi, my dog Alex's skin began to heal. I thought there might be some hope of his getting a coat of hair back. I certainly hoped so for he was at one time a handsome dog and very popular with all of the GI's.

After we got out of the desert many other good things began to happen for us. One of them was the arrival of Eddie Rickenbacker at our camp. He was the first to come from the United States to see us. Col. Rickenbacker was a flyer, the leading ace of WW I, and also a Congressional Medal of Honor winner. While on an inspection tour for the US Air Forces in the Pacific in 1942 his plane was forced down and he was lost at sea on a small raft for three weeks. He came to boost our morale. He told us what a good job we were doing and how much our country needed our efforts. Col. Rickenbacker let us know that

*Capt. Eddie Rickenbacker speaks to the men of the squadron.*

we were special and appreciated back home. He was a very interesting, warm and friendly man who wanted to get to know us better. He spent about a week visiting with different individuals in our outfit. This visit was shortly before our Ploesti raid, and although this was never mentioned by him, or us, I think he must have known that something big was being planned. The men felt that it was a real compliment to have him stay here with us as long as he did.

Soon we started preparations for the big mission. With little said about it, an Engineering Group went into the desert and laid out a complete replica of the refinery at Ploesti. For the first practice run they marked off the boundaries of the refinery with wide white lines made of lime, being careful to avoid the unmarked mine fields the Germans left behind. The first time we flew so low over the target, the props from the planes blew all the white lines away. They were flying only at twenty-five feet and like a bunch of kids playing in the sand, the crews were having lots of fun. The next day the Engineers had to go back and reestablish the boundaries of the target. This time they used cloth banners thinking they would not blow away. The engineers were right, they didn't blow away, but that night the Arabs stole all of the banners just to get the cloth. The engineers had to go back to the drawing board again. This time they marked the target with large drums and five-gallon petrol cans, which were plentiful, and nobody wanted them. They were placed strategically around for identification of the target. This times it worked. Our crews started flying training missions dropping only wooden practice bombs. While the crews were trying to figure out the best way to destroy the target I was busy trying to figure out how we were going to get good, sharp photographs of the target area. Everybody was having fun with these practice missions. The crews liked to kid each other and when they returned to the base they would claim taking out all sorts of things. One crew talked about taking

out the refinery on their run while another crew said they got the storage tanks. Their buddies from another crew raised up and said, "Oh, you didn't do that. We saw what you hit. All you did was to take out three camels and a 'Wog'" (Our name for the Arabs).

Our own Photo Lab Trailer and other American equipment finally arrived. We could now turn out some good quality photos. What a pleasure it was to have at last this photo equipment.

However, a more serious problem arose for the photo crew. Since the planes were flying so low, all of the pictures were blurred. We didn't have film or cameras fast enough to stop the action and give clean, sharp pictures. I flew on several practice runs with the crews to try to figure out how to remedy this problem.

The easy answer was to let the gunners hold large K-20 aerial cameras and take pictures. We tried training the men to use these cameras but this was entirely unsatisfactory and not at all a good option. These men had enough to do without being bothered with cameras. Over the target they would be busy firing their guns and defending their plane. We did however put 12 of these hand held cameras on the planes the day of the Ploesti mission with the instructions camera work was not a priority. They were to take care of their primary duties first but if they saw something they wanted to take pictures of, pick up the camera and use it.

I talked to Col. Kane about the problem of our getting good photos, and he said, "Sgt., this is one mission we must have good photographs. I don't care how you get it done, just get it done. This is perhaps the most important mission we will ever make, and we must be able to document what we do with good photographs."

"Yes, sir," I said, but I didn't have a clue as to how this would be accomplished.

As I began to discuss this problem with our other technicians, I remembered as a kid taking pictures from a train. The pictures I took looking

*Colonel John R. Kane (extreme right) listens as a crew is being debriefed after completion of a mission.*

straight out the window were all blurry, while the ones looking up the track or looking back where we had been were okay. "That is it!" I thought. Here is the answer we have been looking for. Now if we can just figure out how to have stationary mounted cameras pointing straight down to take oblique pictures, we will have it made. The camera mounts would be too difficult to change, but if we could bolt a mirror at the end of the lens, mounted at a forty-five-degree angle, that might do it. At least we thought it was worth a try. The next day I rigged up one camera with a mirror to try on one of our practice runs. When we processed the film, we got excited. This not only worked, but it worked magnificently. We were so pleased about this. The pictures were clear, sharp and reflected where we had been

and what we had done. Now for the next problem. Mirrors in the desert, whoever heard of that? Where could we come up with all of those mirrors?

I was reminded of an often-told true story of Sir Lawrence of Arabia who took several Arab leaders with him to the 1920 World's Fair in Paris, France. When it came time to leave the Hotel, Sir Lawrence went by their room to get them so they could return to their Country. He was astonished to find the Arabs were trying to remove the faucets to take back home with them. The Arab Chiefs told Sir Lawrence how great it would be to have faucets in the desert. I must admit I felt about the same way as the Arabs but my wish was to have "mirrors in the desert."

Throughout this war, the British had not failed to supply needed equipment when we requisitioned it. Once more, we turned to our friends for help. We requisitioned, and they came up with 60 mirrors to meet our needs in the time frame we gave them. Mounting them on our cameras to withstand the vibrations was not an easy task. We made a few more practice runs with our planes and found the cameras mounted with the mirrors worked well. As time grew near for us to hit the real target, we ended our practice runs. On Saturday morning July 31st, the last day before the real mission, the planes took off for the final practice run with live 200-pound bombs. Because of the planned low altitude approach, the bombs were equipped with time delayed fuses. Our make believe refineries in the sand were totally destroyed. No one would ever know what we were bombing in the desert. With the practice mission completed, we were now ready for the real target.

This mission was to be a surprise attack on the enemy. The word "Ploesti was never mentioned by any of us. If we did talk about the mission, it was always "Tidal Wave." We didn't want anything to go wrong with this mission. Although it was a dangerous one we felt good about it because we knew knocking out much of Germany's oil and petrol supply was bound to shorten the war.

The target date, August 1, 1943, was near. The weather in Europe looked good and was holding. All our planes and supplies were in readiness. The mission was on go! The men were on go! Every plane that could fly would be in the air for this mission.

About 24 hours before take off, some kind of a bug swept through the camp and many of the men were left weak with dysentery. This didn't affect the attitude of the combat crews. Many of the ground people who had never flown a mission before freely volunteered to take the place of any man unable to fly. Morale was high. The excitement was high. All of us were eager to see this mission succeed.

The night of the final briefing was a tense night. The crews all gathered in the green hut that was our briefing room to receive the final instructions on the mission. Col. Kane began by saying, "Men this will be the most important mission of your life. It would take an entire army of 10,000 men, many months and perhaps even years to fight their way to Ploesti and smash this target. Casualties would be very high and it would take much too long to accomplish. We are going to take out these oil refineries with 2,000 men in a single day. In doing this you will greatly shorten the time before you and many other American soldiers go home." These words met with a loud roar of approval and applause. "We are going to knock out this target, or die trying." These were not idle, but prophetic words from Kane.

We were told of the flight plans, about the weather conditions for the area, what to expect over the target and how we were to respond. Instructions were also given on what to do if we were shot down. Survival kits were given to each crewmember, and we were instructed how to use them. We were told this would be a 2300-mile round trip and would take 14 or 15 hours of flying time. One bomb bay in each of the planes would be filled with an extra gas tank to provide the fuel needed to make the round trip. Finally, we were told that many of us would return from this mission, but that some would not. We were advised to get our affairs in order and place our personal belongings in the center of our bed when we got up the next morning. It was further suggested that we write a letter to the person we wanted to receive our personal property, and leave it on the bed also. Very sobering thoughts, but I didn't see that it changed any minds about going. There was not the usual excitement, talking, and horseplay as the men filed out of the green hut where our briefing was held. The men were really serious as they went quietly to their tents to prepare for the important mission that was ahead.

It almost seemed like a suicide mission and yet it was not. Every man scheduled to go thought he was the one who would be coming back. All of us felt that way. With all the planning that went into this, every precaution had been made for our return. Even the escape kits contained everything we would need in the event we were shot down. There was a very good silk handkerchief map of the Balkans, money from the different countries, and even a US $20.00 gold coin. Also, there was a first aid kit, food rations, water purification tablets, biscuits, and always the famous "desert chocolate" that was guaranteed not to melt either in your hand or your stomach." Some of the fellows said they knew that this "desert chocolate" was really the same modeling clay they played with as children. The taste was the same, anyway. Perhaps we were kidding ourselves but this survival kit gave us confidence we would be back from this mission.

I was scheduled to fly on this mission with a Lt. Col. Bleyer and his crew. He was the officer second in command to Col. Kane and to fly this mission with him, was my choice. I had flown with his crew before and he was an excellent pilot. This was the crew I wanted to be with. I had flown some of the practice runs with them while we were trying to work out our camera problems. I was comfortable flying with them.

While I was getting everything together in preparation for the early morning take off, Col. Kane sent for me. It was getting late as I found him sitting in a Jeep outside his quarters, alone and looking at the sky. The sky at night in Africa was very dark, but clear, and the stars seemed especially brighter than at home. Col. Kane appeared to be deep in thought and appeared to be going over every detail in his mind of the men and the mission for that morning.

As I walked up, he said to me, "Sergeant, you see those stars up there? They were here long before we came here and they will be here long after we are gone."

I said, "Yes sir, I know."

We talked briefly about this and then he switched to what was really on his mind, the mission ahead. He wanted my assurance that we would have good photographic evidence of the Ploesti mission. I answered, "Yes, Col. we have all the bugs worked out, and all the planes are rigged with cameras. Just for additional insurance, some extra cameras have been placed in a few of the planes. I have confidence that we will get some great pictures." My thoughts about our starting out together in Palestine were racing through my mind. This is the same man who told me, "We didn't come over to take pictures, Sergeant. We are here to fight a war." What a change in attitude!

Then the Colonel said to me, "You are not going on the mission in the morning."

"Yes Sir," I said, thinking he didn't understand the arrangements. "I am flying with Lt. Col. Bleyer's crew."

"I know, but I have changed this," Col. Kane answered. "You are not to go on this one. I can replace a gunner much easier than I can replace you. You are to stay here and to process the film from the planes that do return."

"But Col., I have ground crews who will be here and are trained to do this," I answered.

"We have plenty of men to go on this mission," he continued. "You are the only man I want to process those films. It is that important to the mission. Do you understand these orders?"

I was disappointed, but I said, "Yes Sir," and returned to my tent.

It was almost midnight now. This was an emotional time for all of us. There would not be much sleep anywhere on the base. I went around to the tents of my close friends, to see what I might do for them and also tell them goodbye. There would not be much time for this in the morning. We talked about personal things and how much working and just being together had meant to each of us. The really tough part for me was when friends would ask me to take care of their personal items.

I was correct about there not being much sleep going on that night. Start-

*Aircraft – Down and out.*

*Fires are burning in Ploesti refinery, as another wave of B-24s come in to drop more bombs.*

ing from around 2:00 a.m. and on, you could hear alarms going off all over the camp. The crews were to be up, have breakfast of bacon and eggs at 3:00 a.m. and a final briefing at 4:00 a.m. Spirits were high and everyone was in a good mood, laughing and chatting away about all sorts of unimportant things. The Big Day had finally come.

At the briefing several last minute changes were made on crew assignments. Pilots and co-pilots especially, had to be shifted around to take care of some last minute changes. They had to make sure all planes had a full crew of at least ten men. Some of the crews had an extra man to accommodate an observer.

Just as dawn was beginning to break, the giant fuel trucks rolled out on the runway to cap off the fuel tanks of each plane. The fuel used to check out the engines that morning had to be replaced for at this point every drop of fuel was important.

This was the final task. All was in readiness and the tower control officer made the final decision ending six months of planning. When he fired the pistol with the green flare all was clear to take off. At 6:45 a.m. the first plane rolled down the runway and became airborne. Every 30 or 60 seconds another plane followed until all fifty planes from this base were in the air. They joined another 150 planes from the area to form a great armada of two hundred B-24's loaded to the hilt with fuel and bombs for Ploesti on their mind.

Trouble began as soon as all were airborne. One of the last ships to leave the ground, "Kickapoo," piloted by Robert Nespor, developed engine trouble on take off. Realizing he would never make it, Kickapoo circled out over the Mediterranean and jettisoned their bombs. Our first casualty of the mission came when Kickapoo banked back into the dust to make an emergency landing, clipped a telephone pole and crashed into the desert sands, killing all aboard.

Weather conditions were perfect when our planes crossed the Mediterranean at 2,000 feet, but weather soon became the first enemy as they crossed over the 9,500 feet mountains in Yugoslavia. Here they encountered thick cumulus clouds that restricted visibility and the planes were unable to keep in their formations. Some of them were ahead of where they were supposed to be and got to the target early. When the main flight led by Kane got to the target the only surprise was that another flight had already been there.

Just seven hours after take off the tall stacks of the refinery and their storage tanks were rubble with searing flames and billowing black smoke. Six hundred thousand pounds of high explosive, time delayed bombs, along with hundreds of clusters of incendiaries were dropped by the B-24's flying only twenty-five to fifty feet off the ground. They had been expertly trained in the Libyan Desert to know where the tall stacks were and how to miss them. Every time one of the incendiaries would hit a fuel storage tank, a loud roar erupted, followed by a sheet of flames leaping to the sky.

When the planes came over the target, the peaceful looking countryside suddenly became a living hell. Areas all around the refinery began to open up with anti-aircraft guns blazing. Haystacks and cottages fell apart revealing concealed high-powered guns. Trains even rolled into place with big guns ready to fire. The place bristled with anti-aircraft fire. Our crews had never seen a target so heavily defended. When the second wave of bombers arrived for flight over the target, they found a burning inferno. Still, this did not stop wave after wave of planes as they continued coming over the target dropping their bombs.

Col. Kane led the third flight, the largest number of B-24's, over the most important part of the refinery. When he arrived there, Col. Kane found that because of the weather delay they encountered, some of the planes had already been over the target and dropped their bombs. This was bad news because the bombs which had been dropped had a delayed action fuses and were already exploding. The short fuse bombs were to be dropped last, but Col. Kane did not hesitate. With no concern for his personal safety, he led his men into this exploding, blazing inferno. At a terrible cost of men and equipment Col. Kane's group carried out the mission and dropped their bombs over target.

Lt. Robert Sternfels, pilot of the fourth wave of planes, went into the flaming target yelling, "Here we go!" His plane's wing clipped a balloon cable but he continued on through the flames and black smoke. Their wing mate, Lt. Roy Morgan also got part of his right wing ripped off by a balloon cable. His hy-

draulic system, radio and electric systems were also out of operation. Morgan said, "I figured that closest to the ground was safest." Miraculously, both planes made it back safely.

The fifth and final wave over the target really caught the brunt of the mission. Six planes went in over the target, but only one came out. Lt. Francis Weisler and his crew were the only survivors in this flight. Although his plane was damaged, they made it back to home base.

The returning crews told us about the plane piloted by Lt. Lloyd Hughes and their flight over the target. Gunfire hit their fuel tanks in the bomb bay and also in the left wing before going over the target. Fuel was pouring from the plane in large streams. Flames from the refinery were leaping higher than Lt. Hughes' plane and they caught the streams of fuel pouring from the B-24 on fire. Lt. Hughes could have ditched his plane in a nearby wheat field but elected to continue flight over the target. It was not until Lt. Hughes' bombs were dropped over the target that other crews reported Lt. Hughes attempted an emergency landing of his plane in a riverbed. It was too late, however, as raging fires consumed the plane. Three crewmen out of eleven men survived the crash, but one later died from injuries he received in the crash.

For the men back at Bengasi this had to be the longest day ever! The hours rolled on and we still had no word. Radio silence was not broken until our planes left Rumania. Our total being that day was dominated by our thoughts of the men and their mission. We wondered what was going on, about their safety, and where they were. We had something at stake in this mission also. Then word on the strike finally began to trickle in. Gen. U.G. Ent, Ninth Bomber Chief, radioed from his plane, "Mission successful." The ground crews back at the base let out emotional cheers at this "good news."

Later that night of the Ploesti air raid, the Group Status Board, which was lo-

*Raid on Ploesti.*

cated in the Operations Room of the group headquarters, painted a grim picture. What a devastating toll this mission had been. Out of the 50 planes that took off from Lete that morning, only 13 of them returned to our base. Of those that did return, the planes were so badly damaged and beat up they were unable to fly without major repairs being made.

There were 37 crews that were either "Down over target," "Missing," "Lost," or landed somewhere else. This Status Board told it all, the whole grim story of what had happened. These were not just cold names on the board; they were warm-blooded human beings that represented a crew of ten men that were on the plane. They were our friends, our buddies. We worked with them, we knew them. They were our family. It was hard to look at that board and wonder

what had happened, where the lost ones might be. As Col. Kane said, "We will take out this target or die trying." Well, they took out the target. It was left in shambles, but too many of our men were left dead trying. What a great loss of men and equipment. It will take a long time to build the group back up to strength, and even longer to heal emotions.

Col. Kane's plane, "Hail Columbia," was badly damaged over the target. The number four engine was knocked out and two of the other propellers had holes in them. Because of the bullet holes and the damage to his propellers, Col. Kane knew they didn't have enough fuel to get back to Libya or to cross the mountains. Col. Kane made the decision to head for Cyprus. He was falling back and could not keep up with the other planes when he radioed the others what his intentions were. Three of them advised Col. Kane they were going to Cyprus with him.

Although they were out of the target area and the reach of the fighter planes, Col. Kane and his crew were not out of danger. Col. Kane continued to lose speed and was approaching stall speed when he yelled to his crew to throw out every thing they could. Out the bay windows came guns and ammunition, oxygen tanks and other equipment. One of the pilots in another plane saw this and radioed Kane to ask if they had decided to do a little spring house-cleaning on their way home. At this tense moment when he could neither gain altitude nor speed, the sweat rolled off his forehead into his eyes, and Col. Kane replied in strong language that he didn't think these remarks were a bit funny. They finally got up to 6,500 feet altitude by ditching everything they could, but they needed 7,000 feet to clear the mountains. Their air speed now was only 130 MPH and stall speed was at 125 MPH. Captain John Young tells what happened. "By picking our way through canyons and ravines and some lucky updrafts at just the right time, we managed to get over." When they got to Cyprus Kane had to crash land the plane but all men walked away from it.

I was very glad to see Col. Bleyer and his crew when they landed safely back at our base. This is the plane and crew that I was scheduled to fly with. This was a bright spot for me, on a momentous day. I shared their joy in knowing that they returned home safely.

We spent the rest of the night processing the film that came in from the mission. When our photo lab crew saw the quality of photos we had, we all were very excited. This is what we had worked so hard for and we could hardly contain ourselves. As these dramatic pictures began to develop in the lab where we were working, I knew we had done what Colonel Kane had asked us to do. I told my men, "I can just see these pictures now in *Life Magazine*," and sure enough I did. They were in the August 30, 1943 issue, including a whole spread. Winston Churchill said, "In view of the importance of this target, Ploesti, these must be the most exciting and dramatic pictures to come out of the war in Europe." Apparently many others agreed with him.

Col. Kane and his crew returned to our base a couple of days later. I thought about his words the Saturday night at our briefing: "This is the most important mission of your life." And so it was. I am thankful for the men who pulled it off and lived, and I am thankful for those who didn't return, but died fighting for a democracy they believed in.

This mission was a clear turning point for us. What is ahead for us now? Where do we go from here? Ploesti was one battle won, but the hardest part of the mission for those who were left behind, was yet to be done. Gathering the personal items and letters of the men whom did not return, and sending them to their parents or wives was a tough thing to do. To us these men who would not return were family too. It was difficult. Of those I handled, I think a 22-year-old farm boy from Nebraska pretty well expressed the feelings of all of those who did not return. He said it this way:

*Saturday Night*
*July 31, 1942*
*North Africa*

*Dear Mom & Dad,*
*Here it is Sat. night and I'm sure you are making plans to go to Church Sunday. Say a prayer for me, won't you.*

*We are making final plans for a very important mission in the morning. They say it is the most important one in our life. I want you to know that what I do, I do out of the love for my country. If you get this letter, you will know I didn't make it back. I hope you will understand, and please try to help Bud understand.*

*Know that I love you, and I am grateful for all that you have done for me. Most of all, thank you for the gift of life itself which you gave to me.*

*You're loving son,*
*Ray*

Ray left this note with me, along with his High School ring, a few pictures, a Boy Scout knife, his dog tags and his Flight Jacket with instructions that if he did not return, I should send these to his folks. He wanted his little brother to have his jacket. Sometime after I sent these items to his family, I received a very warm and caring letter from Ray's Mom and Dad. They thanked me for sending the package to them and wanted to know more about their son's life in the military. I answered their letter and told them what a good soldier Ray was, and that they could be very proud of him. I knew they wanted to know how their son died in this mission. I told them something about the attack on the oil refinery at Ploesti, and that all of the men, including Ray, went on this mission willingly. Ray and the other airmen felt this mission would help to end the war. All of us wanted to get it over soon so that we could return home.

They continued to write me and wanted me to come see them when I got back to the States. They said they would pay all my expenses and made other generous offers if I would come see them when I came home. This was very kind of them, but I felt uncomfortable about this and decided I had done all I could for them. I couldn't help them to get their son back, nor did I want to fill his place with them. I decided it was time to cut it off. I wrote back and told them that at this time of the war, I didn't know when, or even if I would be coming back to the States. There was still a war on over here and we had to finish that for which Ray had sacrificed his life. I didn't answer any more of their letters.

For "Conspicuous Gallantry," Col. Kane and also Col. Leo Johnson both received this nation's highest award, the Medal of Honor, for their efforts in the Ploesti air attack. Later the entire 98th Bomb Group received the Presidential Citation for this mission.

Both our crews and equipment were so badly abused from the Ploesti raid, it took us some time to get back in full operation. Twelve days after the big raid we managed to get a few planes up for a mission to Wiener Neustadt. This was to be our first mission after Ploesti. Only 20 miles southwest of Vienna, Austria, the target was a large Messerschmitt aircraft factory, which was heavy defended by both anti-aircraft guns and pursuit planes. It was a very important plant for their war effort. Before we got to the target, weather socked in so solid we could not see the target and had to abort the mission and return to base. It was another week before we could get off a mission to Foggia, an air base in Italy. This time we were able to get up only 18 planes, but successfully did much damage to this base where the Germans ME109's were based. From Lete we put in the air 50 planes for the Ploesti mission. We lost 20 of them and the remaining 30 were left scattered all over the area. When we finally got all planes back to Lete, major repairs needed to be made on most of them.

On September 2nd we tried once more to get a mission off, but we were able to put up only 7 planes. That was enough to deliver much damage to the Railroad marshaling yards at Sulmona, Italy. This was a very important junction on the Northeast coast of Italy. It was supposed to be a regular milk run without much opposition from the enemy. The flight was over water until we went in to the target on the bomb run. This was good news for crews flying this mission, but unfortunately the enemy fighters didn't get the message. The formation was led by Lt. Hoover Edwards who was the pilot on "Sad Sack". Just before our planes made their run over the target, they were attacked by Germans JU-88's equipped with something new rockets! The JU-88's stayed just out of range of the B-24 guns on our planes, and fired their rockets into our formation with devastating results. The number four plane in the formation "Sweaty Betty" took a direct hit, pulled straight up, rolled onto its back and fell away. No parachutes were seen to open and none of the crew was ever heard from.

Lt. Edwards' plane took the next salvo of rockets and his plane was seen going down over land. Five parachutes were observed opening from this falling plane by other crew members. Other bombers went on to drop their bombs on target. Later that night five of the seven planes returned to home base with considerable damage to their planes.

After this September 2nd mission we were back up to strength and hitting the Axis where it hurt the most and with great consistency. For the next three months we continued our bombing raids on Sicily and Italy. The retreating German North Afrika Korps was backed into a corner with no place to go. The American and Allied forces had landed on the other side of North Africa near Oran and were pounding them from the rear. We felt things had really turned around and were now going in our favor.

In September we moved from Lete, Libya to Hergla, Tunisia, a better-equipped airfield for us. The Germans must have been in a real hurry to leave because they left behind so much of their equipment, including a Stuka JU-87 Dive-Bomber. Although, as far as we could tell, it was in top condition, Col. Kane issued orders that no one was to fly this plane. For a couple of days no one even thought about it because of the Colonel's orders, but the curiosity of one of our pilots, Major Hahn, got the better of him. He had always wanted to fly a German plane and see how efficient it was. His curiosity was really no different from the rest of the squadron. We all wanted to know. It was sort of like holding candy out to kids. You know someone is going to take it. Several of the pilots were anxious to try the plane, but it was Major Hahn who couldn't resist the temptation. After looking the dive-bomber over carefully, Major Hahn felt compelled to try it out and disregarded Col. Kane's orders. He took the Stuka up over our field to about 12,500 feet before he started his dive just as though he was going to bomb our field. It was a beautiful sight to see, and very comforting to know the Major did not have any bombs to drop. Pulling the plane out of the dive at 300 feet, he leveled off, circled around and landed without any complications from the aircraft. But he had real problems with Col. Kane. Kane was furious. He asked the ground personnel, "Who is that fool up there flying the German plane?"

They advised him that it was Major Hahn, who just happened to be the Colonel's best friend and confidant. We felt that if anyone could get away with this it would be the Major. Not so. Kane was there in his Jeep when the Major taxied up and killed the engine of his plane.

None of us had ever seen the Colonel this mad before. He chewed out the Major and grounded him from flying anything the next 30 days for disobeying his "no fly" order on the Stuka Bomber. Kane told the Major that he not only put his own life in jeopardy, but also the lives of others. For this flagrant disregard of authority the Colonel told him that he had no place in his Command for someone who disobeyed his orders, and that he would be transferred out of this Bomb Group. In a few days, Major Hahn was gone from our outfit, and we never saw him again.

It amazed us that Col. Kane would come down that hard on someone he thought so much of and who was his best friend. It was a test of power between these two strong men. Kane won out, but of course, he always did.

# CHAPTER SEVEN

## Cairo: The City Revisited - A Dream Realized

Later on in November, I was given a temporary leave of absence from my squadron to go to Cairo for the Air Force to take pictures of an upcoming "important meeting" that was scheduled to be held there. For security reasons they did not tell me who would be here or what this was all about. I was only told to take my Speed Graphic camera and several film packs so that I could get plenty of pictures to record the event.

I was pleased, and somewhat excited, to go on this special assignment for the Air Force. While waiting for the meeting, I stayed at The Shepheard Hotel along with other press people. I was interested to learn that Adolf Hitler had canceled his plans for a great celebration in Cairo. What had once been the hottest rumor in town was now dead. That was not going to happen. No news leaked out about the upcoming meeting I was there to cover, and I was still in the dark about who was coming and where it would be held. Most of the national press corps stayed around the Shepheard Bar and fought the battles there where it was more pleasant to do so, and a lot safer. They would file their stories from Cairo but got much of their information, whether it was correct or not from the bar at the Shepheard.

Two exceptions were reporters Ernie Pyle and Bill Mauldin. They liked to get out in the field and report on ordinary soldiers or airmen. Ernie Pyle, who reported the human side of the war, didn't care about military strategy, campaigns, or battles, but reported on the men involved in them. Pyle was America's favorite correspondent, because he reported on war from the point of view of the men doing the fighting. Bill Mauldin on the other hand was an award winning editorial cartoonist that featured two soldiers, GI Joe and Willie, in all his cartoons. Many of these appeared in the "Stars and Stripes" and were popular with the troops. Mauldin always depicted the average GI as a tired, unshaven, dirty soldier who didn't want to die. He simply drew his characters as he saw them. He knew there wasn't much water for shaving and bathing in the trenches. These were the boys from the Main Streets and farms of America who emerged as gallant, brave, and unpretentious heroic Americans. To discover their feelings Mauldin had to get out with the troops. We had a lot of fun with one of the men in our squadron who looked much like one of Mauldine's characters. We placed him in a typical combat setting, took pictures of him there and sent them to Mauldin with a note that we had found his mythical GI Joe.

One other freelance correspondent and photographer who spent a lot of time at our base of operation was Jim Bray from Birmingham, AL. He worked for the Black Star Agency in New York and covered Middle East War activities. Jim was a cautious and good-natured individual who often went on missions with our crews. This was rare, for most of the national press would not take that risk. Jim came in and out of our operations most of the time we were in the Middle East. He shot both movies and still photographs, and seemed to have the qualities of both Pyle and Mauldin. Upon returning from one of his trips back to the States, Jim brought with him the first color film I had seen. He shared this with me and I began shooting some of our pictures in color. We didn't have

65

much of this new color film so this was a first for both of us.

Cairo was still a great place for soldiers to come and release wartime tensions. There were not very many GI's around, mostly British and Free French soldiers who would come here to get a hot bath and a cold beer. In the evening a cool breeze drifted in across the River Nile and changed the mood of the city. It became a party town untroubled by war. The steaks were tender, the wine was French, and the companionship was friendly. Whatever sort of entertainment you were looking for was there. The highlight of the evening often was a performance by one of the talented belly dancers. The most talented and popular belly dancer in Cairo at that time was suspected of being a spy for the Germans. She was later arrested and convicted of this crime. This neither slowed down nor eliminated the other club activities in Cairo. The "Melody Club" was a rather large club located down town and a very popular place to go for dinner and dancing. I was told they had a very good band but they were set back in an alcove covered by barbed wire to protect the band from brawls that usually broke out between soldiers. I made the decision to get my food and entertainment some place else. The "Melody Club" was not for me.

While waiting for this big meeting there was time for a little sight seeing. I went out to the Pyramids and the Sphinx, which were only about 25 miles

*Winston Churchill wrote in his book, Memoirs of the Second World War, some of his reflections on the Cairo Conference. Before the conference started, he said he took the occasion to visit Generalissimo and his wife at the villa where they were staying. This was the first time Churchill had met Chaing Kai-shek, and he was "impressed by his calm, reserved, and efficient personality." At this time, Chaing Kai-shek stood at the height of his power and fame. He was a strong anti-Communist and the champion of the "New Asia." To American eyes he was one of the dominant forces in the world. Churchill wrote that "in those days, I did not share those excessive estimates of Chaing Kai-shek's power or future helpfulness of China." That evening, he had a very pleasant conversation with Madame Chaing Kai-shek and found her to be "a most remarkable and charming personality." President Roosevelt had all of the participants photographed together at one of the meetings at his villa. Churchill wrote, "Although the Generalissimo and his wife are now regarded as wicked, corrupt reactionaries by many of their former admirers, I am glad to keep this photo as a souvenir." (Courtesy of National Archives)*

*Delegates of the Cairo Conference. (Courtesy of National Archives)*

from downtown. I was very much impressed with these Egyptian wonders of the World. The dynamics and sheer size of these monuments made my mind reel with excitement. How were the builders able to construct them so many years ago without the use of any of the heavy equipment we have today? The Egyptian Government, to protect the head of the Sphinx, had placed sandbags under the chin so if the Germans came rolling through, the head of the Sphinx would be protected. At that time there were few buildings around the Pyramids for they were out in the desert. The Arabs were very proud of the Pyramids and Sphinx and were happy to show and tell us about them. While there I rode a camel and it was unlike any experience I had ever encountered. I had ridden lots of horses back home and I thought the ride would be like that. Not so! It appeared to me the camel's legs were double jointed because when the beast started to get up in one direction and I shifted my weight to balance that direction then suddenly he moved in another direction. This was so unexpected that I would be caught off guard and nearly thrown off. After the thrill of just getting up, the ride was rather uneventful.

I kept waiting and wondering when the meeting I was to photograph would happen and just who was coming to Cairo. Finally on November 22, 1943, we received the news the President of United States, Franklin D. Roosevelt, was in Cairo and would be meeting with the British Prime Minister, Winston Churchill, and Premier Chiang Kai-shek of China to open the Cairo talks. Never in my wildest of dreams had I expected anything like this. Here was a major news story affecting the World, and I was in the middle of it. The meeting was to last four days until November 26.

For security reasons, the Cairo Conference was held out near the Pyramids rather than in Cairo itself. The headquarters for the conference was the Mena House Hotel, just opposite the Pyramids. The American and British Chiefs of Staff all stayed there, and Churchill stayed in a villa a half mile away. Generalissimo Chiang Kai-shek and Madame also stayed a half mile away, while President Roosevelt occupied the spacious villa of the American Amdassador Kirk about three miles down the road to Cairo.

With my knowledge and experience of so many Axis spies hanging around the Shepheard Hotel in the downtown area, I understood why they held this meeting where they did. For security purposes the entire area was isolated and blocked off with barbed wire. The negotiations were held with maximum secrecy. Very little about this meeting was made public. Most of the Press Corps at the Shepheard Hotel were not even aware of the meeting going on here until it was over and the world leaders were safe and out of harm's way.

The talks centered on Far Eastern problems and it was agreed that Japan should be stripped of all her colonies and that China was to have great power status in the postwar world. It was reported to us, General "Vinegar Joe" Stilwell, who accompanied President Roosevelt on this trip, used his influence against Chiang and kept the meetings in a hostile atmosphere. Chiang held his ground, obtaining concessions that he wanted very much from Roosevelt and Churchill. The end results from these four days of negotiations were that Japan was doomed to lose even its oldest colonies, while China would gain in power and world status. With Chiang's strength and persuasiveness, I would say he came out of this conference looking very much like the winner.

The beautiful wife of Chiang Kai-shek came with him on this trip and although she was Wesley-educated and could have participated in the conference, she kept her activities to the many social events of this conference instead of the negotiations. She seemed to enjoy her many shopping trips into Cairo, and would go there nearly every day of the four days she was there. As a member of a B-24 Group, I was interested in noting that Premier Chiang Kai-shek and his wife traveled to Cairo on a B-24 that had been converted to a transport aircraft. This plane was the prototype for the C-87 transport. Apparently Madame Chiang Kai-shek was very comfortable with this plane for she used it as her personal transport on her frequent visits to the United States.

When the three world leaders gathered for the conference, security was really tight. I had never seen anything like this and could easily understand why the Air Force wanted a pool photographer to record the event. I was just glad they choose me to record this meeting.

Before the conference started we were escorted into the room where the conference leaders were to meet. Then we were taken outside to a beautifully landscaped terrace where we were to take our pictures to record this news event. After the conference began we were not allowed to take any more pictures while they were in session. We were instructed that no questions about this event would be permitted and that we were not to discuss this event with anyone. The President was unable to walk and was mobile only with a wheel chair. We were instructed that our photographs were never to show this handicap of the President. I was very careful not to show the wheelchair in any of my pictures. When I finished taking my pictures and was about to leave, one of the aides introduced me to President Roosevelt and Prime Minister Churchill. The President asked me what part of the States I was from.

I said, "Arkansas, Sir."

The President replied, "Oh yes, I have been to Little Rock."

"Yes Sir, I know," I replied. "You are very popular there."

As I left, my thoughts were on another day, June 10, 1936, when President Franklin D. Roosevelt came to Little Rock to celebrate the One Hundredth Anniversary of Arkansas' Statehood. He appeared in Fair Park where there was an auto racetrack and bleachers with lots of seats. A platform had been constructed on the racetrack and equipped with ramps large enough for the President's car to drive on stage. The stage had been constructed especially for him. President Roosevelt addressed the large crowd gathered there and upon leaving, his convertible was driven off the other side of the platform. As his car began to drive off the President saw me standing there waiting to take his picture. He told the driver to stop while he waved and I took his picture. My camera was a "Brownie 116" box camera that probably cost no more than four or five dollars. It was just a step above a pinhole camera. Of course the picture was not sharp. In fact it was not a very good photograph. I simply did not have the right equipment and was very disappointed not to have my personal picture of the President. However, I learned an important lesson in life from this incident. When the President of The United States saw me with a camera, he stopped long enough for me to take a picture. I don't know whether it was because I was a young boy or because I was holding a camera, or perhaps both that caught the President's attention. I thought about the power of a camera and decided right then that I wanted a better camera, and I wanted to learn all that I could about how to use it. This would be my ticket to reach important people and photograph them. Now, several years later, here I am, in a world arena with the leaders who were making decisions that will affect the whole world. I am a part of this event because of a chance meeting between a young boy and a President who never knew the strong influence he had on this young boy. What unbelievable odds that this would ever happen.

With the meeting over, Chiang went back to China, but Roosevelt and Churchill took off in separate planes to go on to Teheran to meet with Stalin from November 28 to December 1. Roosevelt was the first President to travel by air while he was in office. The president left Cairo and flew there in his private plane, "The Sacred Cow."

The President came to this meeting in Cairo in this new plane, a Douglas C-54 aircraft, which had been extensively modified for him. The modifications included a conference room with a large desk and bulletproof window, and with a special feature of an elevator behind the passenger cabin to lift the President in his wheelchair in and out of the plane. This plane carried a crew of seven plus up to fifteen passengers. It was well equipped and modern for its time. For security reasons the ID markings on the President's plane were changed for this flight to Cairo. With all of the intrigue going on here at Cairo, they did not want anyone to know the airplane of the President of the United States was in Cairo.

Afterwards I returned to my base and the 98th Group. Many men in the group knew I had been gone a week, but no one had the slightest idea where I had been or what had happened. We were instructed not to talk about this meeting and who was there, and I didn't. Several knew that I had been away for a week, but just thought I had been to Cairo on R & R. I just let it go at that.

In returning to our base I discovered that the odds were good that we would soon go to Italy, perhaps spend Christmas there. That was sure okay with me and I hoped this rumor was right for a change.

Early in July Premier Benito Mussolini's power and control of Italy were beginning to crumble. By July 25th King Victor Emmanuel told Mussolini that he no longer was head of the government. Since he was forced to resign it was just a matter of time, we thought, until we would be in Italy. However, Mussolini refused to give up his power so he was arrested and placed in prison. It was not long until the Germans rescued him from prison and took him to Northern Italy where the Germans were still in control. Although the Italians had no desire to fight, the Germans occupying their country were fierce fighters and had plenty of fight left in them.

Back in Tunisia, there was 250,000 German and Italian troops bottled up who were forced to surrender near Cape Bon. I have never seen so much equipment, guns, and armament left behind. A friend of mine picked up a German Luger pistol from a pile of arms, and gave it to me as a souvenir. From the North Africa side, the war was pretty well shut down, but for us, it continued in Italy and Southern France. Because of German resistance in Italy, we were unable to go there until later in November so we continued flying our missions from North Africa.

During this period of time the USO found us. There were several celebrities and their USO shows that came to entertain us. The first Hollywood star to tour front-line bases playing for troops was the big mouth comedian Joe E. Brown. He was a stand up comedian who traveled alone and did not have a large group to back him up. He was a very funny man. While he was telling a joke he would stop and point out a GI and shout, "Hey, Soldier, where did you come from?" It didn't make any difference what the serviceman answered. If it was Dallas, St. Louis, or wherever state or city, as soon as the GI answered, Brown would come back and say, "You mean you admit it?" This always broke up the crowd with laughter. He possessed a great deal of empathy for the soldiers in combat because he had lost a son in this war.

Jack Benny brought a large cast and had a big show to entertain us. With him were Frances Langford, Larry Adler, Jack Hailey, and a cast of singers and dancers. They put on a great show that brought much laughter. I don't know whenever I had heard so much laughter coming from these men. When Jack Benny said, "Fellows, the folks back home are having a terrible time with the gasoline rationing back there. They had to revert to washing their clothes with water," he had hooked them, for every man knew what he was talking about. Then when he added, "They can't even get powdered eggs anymore and had to go back to the old fashioned eggs that you have to break," he had them holding their sides. After they finished their show, they spent the next few days visiting the men in the hospitals in the area. The entertainers sometimes did their best performances before one or two, or even a ward full of injured patients. These were emotional times that brought out the best in everyone.

The heavyweight boxing champion, Joe Louis soon followed the Benny show. He was a rather quiet man whom all of the fellows recognized and respected, since he had held this title longer than any other man. He just went around meeting the men, shaking hands and thanking them for the job they were

*Joe Louis, boxer.*

*Francis Langford, singer.*

*Jack Benny and Winnie Shaw, part of a USO troop show in Bengasi.*

*Jack Haley.*

doing. All of the men got a great kick out of meeting Joe. He was a good morale booster.

Bob Hope came to North Africa with his troop in 1943 and followed General Patton into Italy. The first night they were there the Germans bombed a target next to the hotel where Hope was staying. For the safety of Bob Hope and his troop, General Patton ordered the group back to Algiers the next day. I was sorry we did not get to see his show, but that was the way it turned out.

All entertainers in overseas units had to follow strict guidelines of conduct to maintain military security. All USO entertainers on the Foxhole Circuit, as the overseas units were called, were strictly forbidden to reveal their schedule to anyone. This was for their safety as well as for the men they were to entertain.

We felt differently after these USO shows began to come see us. We felt like we did matter after all. How good it was to be recognized. We appreciated what they were doing, for it communicated to us something of the life we once knew and had left behind. It was good to back in touch with the rest of the world, even for just a little while.

It was mid-morning one day when I was in the photo lab preparing to print pictures when I heard someone outside inquiring about me. I stepped out to see who it was. An officer introduced himself as Lt. Clark from the 154th Observation Squadron stationed just east of Oran. He had flown his P-38 plane over to our base to visit his brother for the day. He asked, "Do you know a Sgt. Bill Roseberry with our squadron?"

"Yes sir, I do," I answered, wondering what had happened to my cousin, Bill.

The officer continued, "When I found out my brother was over here in your Bomb Group, I wanted to come see him. Bill was scheduled to come with me today for a visit with you, but he was unable to do so. Since he couldn't come he asked me to give you this package."

I asked how Bill was and he said, "Oh, he's just fine. I think you will get all your answers when you open this package he sent you."

I thanked Lt. Clark and told him that I appreciated the personal delivery. With that he took off to find his brother.

Wrapped in a film box was a long, long letter, which Bill wrote that it had taken him six hours to complete. This was an account of all he and his squadron had been doing since I left them in the States. I was with the 154th my first year in service and it was with them that I received my training in photo reconnaissance work. I was anxious to learn the news since I knew he had been shot by enemy fire, but didn't know any of the details. There was so much news; I had to read his letter again before I could comprehend it all. It was sure good to get the story from his side of the war. I knew almost everyone in his outfit, and it was interesting to learn that we had been chasing the same enemy across North Africa, but from opposite sides of the Continent. Because it had been almost two years since we had seen each other, Bill had a lot of territory to cover in his letter.

I had heard from home that Bill had been shot, but I didn't know any of the details. He filled me in on what had happened.

In his own words, this is what he had to say:

"I was shot on February 20, 1943, some six months after shipping out. We were at a dirt runway field at Houks Le Bain, Algeria, a make shift airfield with light bombers and fighter planes operating in direct support of front line troops 18 miles away at Kasserine Pass.

"I was with one half of the Photo Section in the small town of Youks Le Bain where there was ample water to operate our photo trailer. This was two miles from the airbase.

"The airbase was being bombed and strafed. One plane was separated from the others and was making a wide turn for another run at the field and evidently threw a few rounds from his guns at this small town. With all the antiaircraft noise around us, we didn't know he was firing.

"I was the unlucky recipient of one

of his rounds. It entered the right side of my chest after going through two small notebooks in my right coverall pocket. It came out just below my heart, tearing a three cornered gash in me and my new coveralls.

"In surgery they opened the path of the bullet for about 6" to make absolutely certain it hadn't penetrated the inner wall of my body. The doctors said I was the luckiest soldier in the Army to miss death this close. Their theory was that the notebooks took the thrust out of the bullet and that it probably hit a rib which helped change the direction of the bullet."

Bill was sent to an Army field hospital and was out of commission for about four weeks before returning to his old outfit. The 154th Squadron was stationed only about three hundred miles from me in North Africa, but Bill told me he would soon be shipping out to southern Italy to join the 15th Air Force. They would be doing weather reconnaissance for the heavy bombers. We, too, had been hearing rumors of going to Italy, so perhaps Bill and I will at last get together in Italy.

Our squadron began to get things ready to ship out. We were told that we would go to one of the bases in central Italy that we had earlier bombed. Our photo trailer and much of our equipment would go by ship to Naples and then on down to our new base. We would fly to the air base in Italy when the time was right.

My dog Alex's coat of hair had fully grown out now and was very healthy and shiney. Suddenly everyone began to notice Alex and was again friendly to him. All of the GI's would feed him, rub his beautiful fur, and talk to him. Alex loved the attention but he never forgot who had nurtured him back to health. Alex was a big dog weighing about 40 pounds and whenever he saw me sitting down, he would come running and jump in my lap spilling over on both ends. He and I both had fun with this but there was no way he could fit into my lap. Alex was now one of the GI's and when we went to Italy, he flew with us to the airfield at Brindisi. Colonel William E. Karnes took over the command of the 98th Bomb Group on November 22, 1943 when Colonel Kane was relieved of duty and returned to the States. At this time we were also transferred from the 9th Air Force in Cairo to the 15th Air Force in Europe. Colonel Karnes who replaced Colonel Kane as our Commanding Officer was killed in an aircraft accident in Brindisi in January of 1945. A young West Point officer, Lieutenant Colonel Marshall R. Gray, succeeded his short time as Commanding Officer. It was Colonel Gray who would lead our group through the Italian campaign. We had an inexperienced Commanding Officer and a well seasoned group of men that made for an interesting situation. We all had a curiosity as to how this mix would turn out. Who would be the winner?

# CHAPTER EIGHT

## *Italy: Civilization At Last*

Thank God, we are now out of the desert, and in a land of civilization. Almost everywhere you look, Italy showed signs of a heritage and a way of life that we were fighting for. True, the language was different, but we could relate to their way of life. There was a similar meaning and purpose.

It was on November 18, 1943, that we arrived at Foggia One Airfield in Brindisi, Italy. Our equipment and supplies were coming to Naples, Italy, by ship, and would arrive later. In the mean time, we were to set up the base and make it ready for operations. These were good accommodations for us, far better than anything we have had and we were pleased to be here. We were somewhat familiar with the layout of this base because we had flown over and bombed it a few times.

The Italian people accepted us warmly and were glad to see us here. Some of the Italian planes and pilots were still around the air base when we arrived. The Italian pilots were friendly, showed us around the base, and were helpful in getting everything set up for us. Many Italian civilians were hired by our Bomb Group to provide labor in many of the areas of our operation. The people hired were paid for their work in both money and food, and they were happy to get the work. But the ravages of war had taken a toll on the Italians. Many people, including Italian solders would gather at our every meal and hold out a plate or pan and gratefully accept any scraps left in our mess kits. Nothing was ever wasted. Their search for forage from our garbage cans was a sober reminder to us that all had not been going well here in Italy. It was sad to see how many people were going hungry. It was devastating and inhumane what Mussolini had done to his country and its people. The Italian people were relieved we were there and glad to have us. We were also happy to be there.

I had a chance to ask the question that had bothered me when we flew missions over their country. I asked the Italian pilots, "When we were running bombing missions over your country, how come you never came in to shoot us down?"

One of the pilots answered, "The Germans were also at this base and they insisted that we get our planes in the air, but they couldn't make us attack you."

"Why?" I asked. "This was your homeland. Didn't you want to defend it?"

Several of the pilots joined in. "Yes, we did want to defend our country, but not against the Americans. We like the Americans. Besides we have so many relatives living in America, we couldn't know for sure that we would not be shooting at one of them." This answer satisfied them, but we Americans knew this was a war that the hearts of the Italian people were not in.

All of us were happy to be back in civilization and also enjoyed the change in climate. It just made us feel better. The GI's were not the only ones happy about the move. My dog, Alex, also loved the change. The cooler weather made him frisky, and so he played a little game of frightening the Italians. He would quietly prance up behind an unsuspecting Italian soldier, and then let out his deepest and loudest bark. The Italian would jump about three or four feet into the air and then take off

*When first cokes received in 2 1/2 years, the Photo Crew drank them hot since there was no ice..*

running for his life. I could see the enjoyment and delight that Alex got out of this little trick. Alex would then take off with his head up high and his nose in the air looking for another unsuspecting Italian soldier. He looked like royalty as he pranced off in his search. Alex never once pulled this trick on one of the American GI's. I'm sure the Italians must have smelled different from the Americans because Alex could pick one out every time.

It was not long until our equipment and supplies came in and we were soon back in business. We were at this base only a month before we were transferred to Manduria, Italy. This also proved to be temporary for we moved in less than a month to Lecce, Italy, where we finally found a home base. We parked the Photo Trailer next to a

building and worked out of both locations. This turned out to be one of the best set-ups we had the entire time we were overseas. Before long, we were back in the business of flying bombing missions and fighting the Nazis in Southern France, Austria, and Northern Italy.

It was obvious to us that the Allies had regained control of the seas as our supplies were just now beginning to come in. In addition to clothes and rations, we received our first Coca Cola's. It had been two years since we had seen Cokes and how good it was to have them. Each man was issued one of the traditional 6 oz. glass bottles. We didn't have any way to cool the Cokes, but remembering our fiasco with the beer, nobody cared. We drank them slowly without any ice because we had not forgotten just how good a Coke tasted. This was such a great treat for us.

The men of our squadron now seemed to be better adjusted and without as many emotional problems. I no longer had to punch nearly as many "Tough Luck Cards," nor were there as many "Dear John" letters from home. We didn't seem to worry as much about getting back home, or personal safety. Sure we were concerned, but we didn't worry about it. We just accepted our responsibility because the job had to be done, and that is what we came over here for. "Let's get on with it, and get it over with" was the attitude. We had bonded as a strong unit and proved that we could take on whatever came along. When bombs were dropped out of our planes, there was no thought of killing individuals. We were disconnected from any such thoughts. Our mission was to take out a plant, a railhead, a harbor, or an airfield. These were the targets of our mission, and targets to be destroyed were impersonal. We didn't think about the fact we were killing people.

It is December again and we are certain to spend yet another Christmas away from home. This will be my second Christmas overseas, with each one spent in a different country. Each time I tried to create some semblance of Christmas back home. I made decorations but they were neither elaborate nor fancy. In fact, my Christmas in the desert was pretty sad. There was not a tree within fifteen hundred miles of us, but I was determined to have a tree for Christmas. I looked all around and couldn't find anything in the desert that even slightly resembled a tree. The best I could do that year was to fashion a tree out of wire. It didn't look like the Douglas Fir we set up in Arkansas but everyone got the idea. I decorated this "tree" with foil from some film wrapping we had already used, and anything else the fellows felt would be appropriate. Some of the GI's came by our photo section to add something to our "tree." On the tree they placed pictures, medals, bright and shiny buttons, and even a star that one of the men made. We

*Holiday greetings.*

may have been two thousand miles from home, but this brought warmth to our hearts together with memories of past Christmases. Everyone enjoyed the sad "tree" in the desert so much that I wanted to have a real one this year.

Our Photo Lab crew was excited about Christmas in Italy. Nearby was a village where we could go shopping for items to decorate our tree. Of course we didn't expect to find the decorations we were accustomed to seeing back home, and we didn't. We were not disappointed, however, for when we made the trip to the village, we found some interesting and beautiful things. The GI's with me were surprised to find live turkeys in the village market. A couple of our men wanted to buy one and take it back to the base to cook for Christmas dinner. I told them, "No, No, we can't handle that. It sounds great, but believe me, it won't work." I think they may have been halfway serious about the turkey. It was difficult but I talked them out of buying a live turkey to take back to camp with us. We found an Angel and some very colorful fruit to decorate our tree. It looked great. All of the GI's enjoyed our Christmas tree and many came by to see it. This small gesture to remember Christmas made me feel close to home for a short time.

When we were shopping in Brindisi, we learned the local Opera House had American movie films on Thursday, Friday and Saturday. We were pleased to learn this for we had not seen many films since leaving home. We didn't have time to go on this trip, but we knew we would be back to see a movie film made in Hollywood. This would be a treat we would look forward to.

Our new Commanding Officer, Lt. Col. Marshall Gray, was a recent West Point graduate who came to us from the States and had no combat experience. He was a tall Texan, who felt everything should be run by the book, just as he was taught. As our Commanding Officer it didn't take him long to assess our situation and he didn't like what he saw. Lt. Col. Gray called for the Headquarters Squadron to assemble on the flight apron at 1400 Hours. All of the officers and enlisted men assembled there had to stand at attention while Col. Gray dressed us down. He told us how raunchy we looked, and that our behavior was even worse. He let us know that he would not tolerate such action from us. In the future we were to salute the officers in a snappy manner and answer with a "Yes, sir" or "No, sir." We were to wear the uniform in a proper manner only, and it was to be clean at all times. He ordered us to get haircuts and to clean up our act. Col. Gray wanted the squadron to project a respectable image at all times.

Well, we got the picture. Lt. Col. Gray thought he was still back at a base in the States, and wanted his outfit to look and act like it. He had no concept of what we had gone through the past three years and how the officers and enlisted men had bonded during this time. We were a unit all right, stronger and tougher than he knew. We had been tested and molded by the fire of combat.

Later that same night, the officers got together with Lt. Col. Gray in the Officer's Club, where, I was told, the drinks were plentiful and stiff. As the night grew late, the drinks became stronger and came around more often.

The officers kept telling Gray how wrong he was in judging the men. Finally he began to see the error of his ways. When he left with a group of officers who escorted him to his tent, they discovered a fire in his tent. No one knows how such an unfortunate thing happened. The base fire truck had to come put out the fire but not before all of Lt. Col. Gray's clothes were wet, and jumbled into a terrible mess to prevent them from catching on fire.

Gray apologized to the men around him that night, but he said that was not enough. He wanted another squadron assembly called the next day. His remarks to the entire group were shorter this time. The first words he said were "At ease men." The day before we stood at attention the entire time he spoke. We saw this time his attitude was different. He apologized to all of the men, and said they looked just fine to him. He was proud to be the Commanding Officer of such an outstanding group of men and respected the record we had. He acknowledged what we were doing before had been successful, and that was good enough for him. He stated we were to carry on as we had been doing for the past three years. When Col. Gray dismissed us, there was a loud roar of approval from the men.

Some may think this was a cruel and unkind thing to do to our new Commanding Officer, and perhaps it was but you must remember that the age of those men, including the officers, ran from eighteen to twenty four years old. Kids, just kids, and yet, seasoned men with heavy responsibilities. We were not accustomed to taking flak anywhere but over the target, and only from the enemy's antiaircraft guns. We did not intend to take any other kind of flak. Obviously we didn't have any more trouble from Col. Gray. As time went on, he became more popular with the men and a fine Commanding Officer.

This little episode made us more aware of who we were and the way we looked, and at the same time Lt. Col. Gray gained a healthy respect for what our men had been through and what

turn to her native Germany and entertain their troops. Ms. Dietrich refused to do so for she had a sister in a German concentration camp. Adolf Hitler then placed Marlene Dietrich on his infamous death list. She was a very well known lady who had some close calls on her life but she never quit entertaining the troops. During the battle of the Bulge Dietrich and her group were about to be captured by the German troops when they were rescued by soldiers from the U.S. 82nd Airborne Division. She never let this stop her. Marline Dietrich gave a wonderful show in Italy that boosted the morale for the battle weary homesick soldiers. They enjoyed her show and kept cheering, "Just one more encore, please." She was a great lady who put on a great show. The men loved it.

After I left the 154th Squadron in the United States, our groups were worlds apart and now I was only 70 miles from where the 154th was stationed. Since this was my old outfit they invited me to come to Bari and have a reunion dinner with them, an event which was something to write home about. My friend, Sgt. Forrest "Buddy" Diemer always had an eye for good food. Back in the States, "Buddy" could always tell you where the best food in town was located. He still maintained that talent and even in Bari, Italy, he knew where to go for some really fine Italian cooking. He arranged for twelve of us to go to the home of "Mamma" Stopans where she had prepared a seven-course gourmet dinner for us. She had a large dinner table that seated all twelve of us. It was set with white linen, beautiful china, crystal, silverware, and fresh cut flowers. You knew immediately it was going to be a great meal. First came the Antipasto, marinated mushrooms with thinly sliced zucchini, green onions, olives, and Grissini (long thin bread sticks). Next came the soup, a cup of minestrone loaded with fresh vegetables, and after that fettuccine with cheese and cream. The main dish was Pollo alla Cacciatore served with generous portions of spaghetti seasoned

only with butter and garlic. A vegetable lasagna with thick tomato and basil red sauce on the side was served with this dish. For desert we had rich Italian wine custard, called zabaglione, which was served over fresh berries with almond cookies on the side. After two hours of eating and talking, we were stuffed. "Mamma" Stopana then came around with some very small glasses and an orange liquor for everyone. I thanked her but told her I couldn't hold another drop, and besides, I didn't drink. She said to me, "Nonsense, you don't drink this to get drunk. You drink this to aid the digestive system, so you don't feel stuffed." I accepted her offer and she was right. What a wonderful way to top off a great evening. It was not only good food, but it was good to see a group of old friends once again after three and a half years.

My Lab Crew had never given up the idea of returning to Brindisi to go the town Opera House and see a movie. We loaded up a crew one afternoon and went to town and discovered the film was an old John Wayne cowboy movie with an Italian language sound track and English sub titles. We thought this ought to be a riot watching John Wayne speak in Italian. When I stepped up to the box office to buy my ticket the girl selling them spoke no English, and I had some difficulty understanding her. It was beginning to create a minor commotion when a young Italian man who did speak English stepped up.

"Pardon me sir, perhaps I can help you."

"Please do," I said.

"The young lady wants to know which price ticket you want. When you purchase your ticket here, there are two prices. One is general admission for the downstairs seats and the other is for box seats with a woman for your pleasure. She wants to know if you want a girl with your ticket."

I thanked the young man, and said, "Well, no wonder I was confused. That's not the way we do it back home."

The GIs with me on that day bought the general admission seats without the girls. I can't vouch for what may have happened on other trips to the movies, but on this trip, we watched the movie and other people. When the movie began we laughed at hearing John Wayne speak Italian in a rather high pitched fast voice that sounded nothing like him. It was a hoot! The movie was divided with two intermissions. A loud bell rang long enough to notify people that the lights were about to be turned on, so they could make themselves presentable. Some of the people had trouble making the five minute warning the bell gave and were still arranging their clothes when the lights came on. The lights stayed on for about thirty minutes giving us plenty of time to buy refreshments. It was amusing. Some of the patrons looked as if they really needed that break while others could hardly wait for the lights to go out again. This was a lesson in life as we saw how other people lived.

The women in Italy were beautiful, friendly, and many of them were easily available to the American GIs. Some of them too much so, for they were ready to trade sex for a Hershey bar or a sack of sugar. The rate of exchange was two candy bars or 5lbs of sugar. When you would see one of the GIs go out at night loaded with candy bars or a sack of sugar, you knew what he had on his mind. The older Italians deeply resented this and would take matters into their own hands. Many of these loose women had their hair cut off and heads shaved by those who resented what they were doing. Many of our American troops become infected with a venereal disease transmitted by these women of the street. Our Medical Officer became quite concerned about this health issue and felt that our men needed a better understanding of this problem. In cooperation with the local Police our Medical Officer decided it would be a good idea to take pictures of girls who were known to be infected with a venereal disease and post these pictures at our base for all men to see. The Doctor asked me to go with him to the police station and take photos of

these girls. At first I refused. I felt really bad about this and didn't want to take the pictures. It appeared to me it was an invasion of the girls' privacy. I recalled Col. Kane's words: "We didn't come over here to take pictures; we came to fight a war."

When I told the Officer that I thought it would be an invasion of the girls' rights, he responded, "Sgt., these girls do not have a right to infect our men with this disease." He further explained to me that this effort could save our men some very serious medical problems, and would help to keep this disease from becoming an epidemic, I then became willing to take the pictures and post them around our base camp. It did help slow this problem down and made the men more aware of the seriousness of the situation. This is one problem we did not have in the desert of North Africa because there were no women, at least none that interested our men.

I have often referred to the "men" in our group. At the time I went overseas with the 98th Bomb Group, there were no women in our group. We didn't see any American girls until we got to Rome and then they were American Red Cross workers. There was a large USO Canteen in Rome that was staffed by Red Cross girls who were there to serve coffee and doughnuts to all the men in service. Sometimes the USO had a band to play music for entertainment and for dancing. What a welcome relief this was for a war weary bunch of American GI's. It was like a touch of home in a foreign land. It was appreciated by all of us and provided a respite we all needed.

One piece of equipment in our photo trailer was giving us a problem and needed repairing. One day I was lying on my back on the floor of the trailer working on the equipment when I heard a voice say, "Sergeant, why don't you get your butt off the floor and let someone who knows what he is doing get in there and fix that equipment."

I was shocked that anyone would talk to me like that. I could not imagine who would say this to me. I came up about to give somebody a piece of my mind, when I saw that it was my brother-in-law, Ken Adams, who was an officer in the Seabees!

I shouted, "Ken, what in the world are you doing here?"

"I came to see you."

He was one of my favorite people in the entire world and when I last saw him, he was constructing Army bases for the U.S. Army in the States. My cousin, Bill Roseberry and I had at one time worked for him. When the War broke out in December of 1941, Bill and I were already in service. Ken thought he should be in service also and went down to join the Army. The Army turned him down because he had a contract to build Army bases throughout the South. The officials told him that what he was doing was more essential to the war effort. For the next two years

Ken continued to build bases for the Army, but he said he still did not feel good about Bill and I being overseas fighting the war for him. He wanted to be a part of the war and do his own fighting. Try as hard as he could, Ken was unable to get into any branch of the service. After soliciting the aid of a very strong and powerful United States Senator, Ken began to make some progress in his quest to be in the service. Ken received an appointment as a commissioned officer in the Sea-bees which was the Construction Battalion of the Civil Engineer Corps of the U.S. Navy. They built harbor facilities and airfields all over the world for the Army and Navy.

As far as I knew, Ken was still in the States. You can imagine how surprised I was to see him. With all that effort just

*Ken Adams came by to look up his brother-in-law, Tech. Sgt. Blundell..*

*Shows total destruction of the village of Anzio and the monastery on top of the hill.*

to get into service, I still did not understand how he found out where I was stationed and how he ever found me. We had a good long visit that seemed like old times. It brought back to me the reality of Little Rock and the family and friends I left behind so long ago.

Bill Roseberry's squadron had now moved to Italy and was stationed over at the sea coast town of Barie, only about seventy five miles from where I was stationed. I told Ken about this and he took off to look up Bill.

The lure of seeing the City of Rome and the Vatican City was strong for many of us. Arrangements made by Pope Pius XII with Hitler and Roosevelt left these cities untouched by the ravages of war. At the time the city of London was being devastated by German bombs and rockets, both sides agreed

with the Pope to leave Rome and the Vatican City untouched by gunfire. This was a tremendous achievement by Pope Pius XII, but then he was a remarkable man and a distinguished statesman. We were anxious to visit the city saved from war by the Pope.

Some of my friends from the photo section went with me on a three-day pass to Rome. The purpose of the trip was to visit and photograph this beautiful city of almost three million people. Just thirty miles south of Rome was the resort town of Anzio Beach which we had to go through before we got to Rome. Anzio was a sea port village where the Germans tied down the Allied forces for four months before being able to break through at Cassino. Going through this area we saw the worst destruction we had ever seen. The houses, the trees, everything was totally destroyed. We saw how devastating war could be, and it was very ugly. The Allies mounted a major attack force and then went on to free the city of Rome from German control. On the morning of June 5, 1943 soldiers of the 5th Army entered the city, bringing it liberation and not conquest.

Rome was occupied by the Germans in September, 1942 and it was spared the ravages of war that fell fate to her sister city of Naples when they left on June 4, 1943. We know from the records that the Germans did not save the City of Rome out of a deep respect or love for the city. It was out of fear at bringing down the wrath of the world on them that caused the Germans to leave the city virtually unharmed when they left.

When we arrived in Rome we were much impressed by the size of the city and its beauty. I remembered seeing this great city from twenty four thousand feet on one of our bombing missions to Naples. Of course our target was not the City of Rome, but it was as impressive then from the air as it was now. Being here on the ground was awesome. Everywhere I looked I saw so much of history spread out before me.

What an opportunity we had to absorb the culture and antiquity of the Eternal City. At the very sound of the name of Rome what thoughts are aroused in each of us? Perhaps it is the almost forgotten lesson of Brutus stabbing Caesar, or Mark Anthony proclaiming, "Friends, Romans, countrymen, lend me your ears," or of Nero playing his fiddle while the city burned. Or one

*More destruction to sea-coast village of Anzio.*

might think of the grandeur that was Rome's, and of the mighty legions spreading their empire throughout the world. Whatever thought may come to our mind at the sound of the name Rome, we think of her as being more than an Italian City. Rome belongs to the world because people everywhere have attached themselves to its common heritage. Rome is truly the fountain of civilization.

The building, which made the greatest impression on me, was Rome's Coliseum. Still standing were enough of the ruins to grasp the size and design of this impressive stadium that in its time could seat fifty thousand spectators. It was opened the 21st day of April in 80 AD with games and contest that lasted 100 days, during which thousands of men and beasts lost their lives. To know and see this place where early Christians were persecuted was quite moving. But in Rome, everywhere you looked you saw a part of history before you.

In Rome, we found two good examples of modern structures that stood in contrast to the antiquity of this great city. The Victor Emmanuel monument was big, bold and dominated the center of the city. Made out of gleaming white Italian marble and decorated with gilt-bronze statuary... it was striking. The other modern monument was the Foro Mussolini built in 1931-33, as a tremendous center for the physical and political training of Italian youth. Its most striking feature was the stadium which had 60 larger than life statues of athletes carved in marble standing on pedestals around the field. This was

*Saint Peter's in the Vatican City.*

*View of the Vatican City from the top of Saint Peter's Cathedral.*

very impressive for they looked as if they were Greek gods watching over the field.

Since the fall of the Roman Empire in the fourth century, Italy has been divided into numerous principalities. In the royal families of Italy there were three kings who were named Victor Emmanuel. During the reign of Emmanuel II, in 1860, a movement for unification for Italy began. It was at this time, the guide told us, that the Vatican was excluded and became a tiny sovereign state situated entirely within the city of Rome. The Vatican City is a complex with historic buildings filled with artworks.

The most imposing and beautiful building in Vatican City is Saint Peter's Basilica. It is the world center of Roman Catholic worship and was designed by the most famous artist and architects of the Renaissance period, including the well-known Michelangelo who designed the dome. This building and Saint Peter's Square, which is part of this building, is so impressive it is overwhelming. We spent most of a day just taking pictures of and in touring the building. Although they would not let us take pictures inside, we did go in to see a most outstanding collection of art. It was unforgettable in its splendor and was placed there to remind us of the beauty of God.

On this trip, we learned that the Pope took some time to have an audience with the American troops who were in Rome. I found out the dates and time so that I could come back on another trip for an audience with the Pope. Although I was not a member of the

Catholic Church, I knew enough about the Pope that I wanted to meet him and be a part of that audience.

Pope Pius XII was born in the city of Rome to a distinguished family. Born Eugenio Pacelli, his father was an attorney for the Vatican. After Eugenio completed his education, his father expected him to join his law practice. Instead Eugenio decided to become a priest. Before being elected Pope in March of 1939, Eugenio served under four other Popes. This experience gave him a great depth of knowledge into the affairs of Europe and some of their problems. He traveled extensively and was the first Pope to visit the United States. He spent enough time in the United States to gain a good knowledge and understanding of our country and its people.

Soon after he was elected Pope in 1939, Pius XII condemned the belief in racial supremacy, totalitarian governments, and the growth of materialism in the world. He pleaded for a "living wage" for all workers and defended the workers' right to organize. He pleaded for world peace, and in these efforts he consulted with President Roosevelt's personal representative at the Vatican as well as many other diplomats. Later when the German armies occupied Rome in 1943, he was successful in keeping both Nazis and Allies from destroying Rome and the Vatican City. This in itself was a brilliant success and demonstrated the Pope's diplomatic skills.

I had seen enough of the City on this trip and heard enough about this man to know that I must come back again for more. I managed to schedule a time when I could come back and have an audience with the Pope.

And so it was back to business at the base. Since our move to Italy, we were still hitting targets in Europe. We were stationed in Brindisi for a short stay then went to Manduria, Italy. In January of 1944 we moved to Leece, Italy where we stayed the rest of our time flying missions to Europe.

If the ground crews ever got tired of getting the planes ready to fly, or the combat crews got tired of flying missions over an enemy target, it didn't show in their attitude. One reason was that the men always kept a keen sense of humor about themselves. Col. Marshall Gray was our Commanding Officer at the time of our tour of duty. This poster which poked fun at us showed up in many of the buildings around the base at Leece:

---

*Gray's Continental Air Lines*
Better buy your ticket tonight. Make your reservations early. Daily flights to Schwekeleschtein, uscherskunken, and Grubbelstahger. See the Vaderland from the air.

Don't miss this opportunity to see the single-engined, twin-engined, four-engined and 12 engined products of German industry. See the beautiful pattern that nine million precision flack guns can put in the sky. Trips? Daily, every hour on the hour. If you find today's plane crowded get your tickets for tomorrow's trip.

Bring along your own oxygen mask, flak shoes and short snoter. *SEE YOUR OPERATIONS OFFICER TODAY*

---

This kind of humor greatly relieved the tension felt by our combat crews. Life among these men was not grim, but the tension of their day was great. As the day goes on, this tension continues to build until they become airborne. Once they return to base and they get their feet back on the ground their feeling of relief is tremendous. Off to bed by ten or eleven o'clock, a good night's rest and they are ready to go again. The cycle repeats itself.

Although we were hitting new targets now, with some new planes and new crews, the attitude which brought us through the desert, was still very much the same. We were hitting high priority targets in Germany, Austria, Hungary, Romania, Southern France, and Northern Italy. Many important targets were made inoperative by these

*Author receives the Bronze Star in Italy from Col. Marshall Gray.*

bomb strikes. As a group of American GI's, we were friendly, fun loving, and good natured; but when it came to our work, we approached it very seriously with an unselfish devotion to duty.

The men of the 98th were very proud when we received the Presidential Citation in March of 1944 for the daring raid on Ploesti back in August of 1943. And then, which made it even better, the next month we received another Presidential Citation for our support of the 8th Army in North Africa. As they say, this doubled our pleasure. For this occasion General Nathan Twining, Commanding General from the 15th Air Force, came to make the presentation and awarded the Banner with the Flag to our Commanding Officer.

All the same, I was still mindful of my visit to Rome and was anxious to go back to have an audience with the Pope. I put in for a three-day pass and hitched a ride with some fellows from my base going to Rome. I had arranged this trip with my cousin, Bill Roseberry, so he

*Airmen from the 98th Bomb Group stop at some of the old ruins of Rome.*

could be there at the same time. We had plenty of time before our appointment to see the Pope, so we went right on out to the Vatican to look around. There was much to see; we could hardly take it all in. My previous trip helped me to know how best to spend our time there but it was still difficult to take it all in.

Our guide gave each of us a copy of *A Soldier's Guide To Rome* and told us the Vatican City covers 109 acres and is an independent state under the authority of the Pope of the Roman Catholic Church. It is located entirely within the city of Rome, Italy, and is the smallest independent country in the world. He said they even have their own currency and postal system, their own telephone service and newspaper and a radio station. Located in the northwest part of Rome, Vatican City is just west of the Tiber River and is surrounded by medieval and renaissance walls with six gates of entry. We saw the Swiss guards who maintain security and protection of the Pope.

The most impressive in the group of buildings were Saint Peter's Basilica and Saint Peter's Piazza in front of the building. This imposing edifice was built between the 15th and 17th centuries and was designed by many famous artists of that period including Bramante and Michelangelo. The other major building in the Vatican was the Papal Palace, a complex of buildings containing more than a thousand rooms housing the papal apartments, the government offices, several chapels and museums, a vast library, and the Vatican gardens. The Vatican Library is a work of art in itself, containing some of the most valuable ancient manuscripts in the world. We were told there are over one million bound volumes of books and manuscripts to be found in the library. The tapestry exhibit in the Museum at the Vatican is only one of the outstanding examples of fine art displayed there. The Museum of Egyptian Art, from prehistoric times to the 6th Century, contains fascinating exhibits. The most famous part of the Palace is the Sistine Chapel with its great ceiling frescoes painted by Michelangelo. He painted these vaulted ceilings over a period of four years on a scaffold 60 feet above the floor and much of the time while lying on his back in a tight cramped position. He painted hundreds of giant figures that made up his vision of the creation of the world. The most famous one is "Adam," where God reaches out and touches the finger of man. The nine main scenes depict Bible history from the creation to the flood. More than twenty years after Michelangelo completed the fresco for the Sistine Chapel, he began his enormous fresco *The Last Judgement* which is a vast painting covering the entire wall of the Chapel behind the altar. It's size, plus its bold, daring conception and excellent presentation makes it one of the truly master works of art on display there. In the picture gallery alone, my head was reeling from the literally hundreds of paintings of the masters including Leonardo Da Vinci, Raphael, and Michelangelo. We saw *David* the almost life size statue in bronze by Donatello, the Michelangelo statue of David, Moses, Julius Caesar, Jeremiah, Micah, and perhaps the most famous of all his statues, the *Madonna and Pieta*, where Mary is holding the dead Christ in her arms.

The sheer magnitude of all of this was overwhelming. We saw the battlefield at El Alamein as a legacy of loss. Now in contrast, the beauty of this place stood like a legacy of love. The great minds of men of this universe assembled here the greatest products their minds, their hands, and their hearts produced over centuries of time. Here we saw man's tribute to God, and recognized what a miracle all of this escaped the destruction of war.

It was almost time for our audience

*Elaborately decorated altar inside Saint Peter's Cathedral.*

with Pope Pius XI, so we assembled in St. Peter's Basilica to await his appearance. There were forty or fifty people waiting, most of them American GI's with a few sprinkling of soldiers from other countries and a few Italians. Bill Roseberry and I bought six rosaries apiece in Rome before we entered Vatican City. These were to be blessed by the Pope. Neither of us was Catholic, but we thought this would be meaningful to our friends back at the base. We were told by the attendants that the Pope would do this for us. As he came into St. Peter's Basilica, the Pope was magnificent in all his robes and accouterments. The attendants that walked the slow cadence with him were chanting in Latin. The darkened room with the bright skylight beaming in from above, added to the holiness of the moment. The Pope reached our group and started giving blessings to us in English. I held out in both hands the rosaries that I had bought. When he came to me, he reached out with his left hand and blessed them and placed his right hand on my head and said, "God's blessings upon you, my son."

When he had finished he sat down and delivered a short address to us. He talked about the goodness of God and how God created man in goodness, and expected us to have love for our fellow man. When we are harmful or disrespectful to our brothers and sisters, then we are harmful and disrespectful to God. We should therefore show love for each other and for the God who gave us life.

It had been a long and full day. I left St. Peter's greatly inspired by all the art we had seen and the audience we had with the Pope. As we were leaving, I turned to Sgt. Bill Roseberry and said, "Bill, with the enormity of this event it is difficult for us to assess the full impact this will have on us. But do you realize that today two Southern, Methodist boys from Little Rock, Arkansas, had an audience with the Catholic Pope in the Vatican City in Rome and were blessed by Pope Pius XII. Just think of all the connections from on high the Pope must have."

That night, at the Villa where we were staying I had much to reflect upon and much about which to be thankful. I thought nothing could surely top this but the next morning I had a great surprise awaiting me!

*View of the ancient Roman Colosseum where games were held and the Christians were persecuted. It would seat 50,000 specators.*

# Chapter Nine

## U S A : *Home Again*

The next morning after our visit to Vatican City, I got up to have a leisurely breakfast with Bill and to make plans for some more sight seeing that day. It was not difficult for both of us to become spoiled by the Italian way of eating meals. We ate a little and talked a lot between courses. It was so relaxing, and enjoyable neither of us was in a hurry to move. I think the afterglow from yesterday's meeting with the Pope was still very much on our minds and spirits. As we sat we talked about the events that had happened the day before.

Seated at an outdoor table at the Villa where we spent the night, we could see and hear everything going on around us. I noticed two Military Police as they walked into the Garden Cafe and stated they were looking for a Tech. Sergeant Blundell. I motioned for them to come over to our table and said, "I am Sergeant Blundell. What is the problem?"

"Oh, no problem, Sir," he replied. "It's just that the Commanding Officer from the 98th Bomb Group asked us to find you and bring you back to the base immediately."

"But I still have another day left on my pass. Here I will show it to you."

"I know your pass is legal, okay," he told me, "but the Air Force wants to send you back home to the United States. That is what I'm trying to tell you."

"What! I can't believe what you are saying is true," I said.

"Believe me it is true!" the MP replied. "The Army has developed a point system for sending men back to the good old USA, and I was instructed to locate you and bring you back to your base as soon as possible. You are to return to the States in the morning."

I told the two MP's it wouldn't take me but five minutes to grab my clothes and I would be ready to go. I grabbed what I thought was mine, told Bill a quick goodbye and to take anything I may have left behind and that I would see him back in the States. We took off in a hurry.

As soon as we returned to the base at Leece, the two MP's took me to the Officer of the Day and the Operations Officer. They sent for Colonel Manzo and then they proceeded to tell me what was going on. The Operations Officer told me the Army had just established a rotation system for returning men who had been in a combat area for a long time back to the United States. Points were given to the men based on length of time overseas, number of combat missions flown, the battles and campaigns they had been in, and some other considerations. It had been determined that I had the highest number of points of any soldier in the European Theater, so I was told, and I was ready to be returned to the States.

By this time our Commanding Officer, Colonel Manzo, had arrived. He congratulated me; told me what a good soldier I had been and that I had made valuable contributions to the Group during the time they had been overseas. Then he said, "However, if you elect to stay here with this Group and not return to the States at this time, I am authorized to offer you a Direct Field Commission as a Lieutenant. We would hope you will stay with us, but the decision is yours."

"Colonel, Sir, I really appreciate the offer, but since I have been through ten campaigns, flown over a dozen bombing missions, and I have been in a combat zone for two years, six months and fourteen days, I feel that I have pressed my luck far enough. I would really like to go home. I hope sir, you can understand my feelings about this."

"Yes, Sergeant," The Colonel replied, "That's about what I thought you would say. I completely understand. Good luck to you in the future and may you have a good time back in the States."

The Army had decided to use me as an example to show the rest of the men left behind just how fast this new rotation plan would work. I would be in New York within 24 hours, and was to leave on a plane for the USA in the morning. My head was reeling. I couldn't believe what I was hearing. At last, I would be homeward bound.

Headquarters was anxious for this word to get around for there would soon be others to follow on this new rotation plan. It was thought this would boost morale among all the men. This news did spread fast: "JEB is going home." It seemed as if everybody was coming around to say "Goodbye," and many of the men gave me phone numbers and asked me when I got back to the States to call a wife or sweetheart or Mother and give them an update and tell them they would soon be coming home. It was hard to think about what I wanted to take back with me, because time was so short, and there were so many interruptions. Not that I minded the interruptions, for they were all well wishes. I really wanted to keep my .45 pistol which I had worn so long and kept by my side all the time I was overseas but was not permitted to do so. It was sort of like giving up a true and trusted friend, but I had to do it. My .45 was one of the first things I had to check in along with my M-1 rifle and other items of clothing I had been issued. I planned to leave Italy the same way I came over here, with very little clothing and as few personal possessions as I could manage.

Amidst all this excitement, I suddenly wondered, what will I do about my dog, Alex? Surely I won't be able to take him with me, but I can ask. I didn't want to leave him here because I knew how much he disliked the Italians. He has been spoiled by the GI's in our outfit and thought he was an All-American dog. Alex was a very smart dog and somehow he knew that something different was going on that night. He didn't want to leave my side, and stayed close to me. If I sat down, he wanted to be in my lap, and Alex was too big for that. I did ask about taking him with me and was told that I could not do that. Several of the men said they would look after Alex, but one of the men in our Photo Section who was very fond of Alex and had helped groom the dog, agreed to take care of him. I told him I would check to see how we might ship Alex back to the States and would be back in touch with instructions. We were to fly out early the next morning and as I went to breakfast that morning, Alex went with me. When I told my dog "Goodbye," I knew I would never see Alex again. We both were sad about this.

We took off early that morning to return to the States in one of the older B-24's scheduled to be returned to the United States. The regular combat crew and I were the only ones on the plane. I was the first person from our Group who was being sent back on the new rotation plan. It was a routine and rather dull flight. We were not loaded with bombs and were not expecting any flak from antiaircraft guns on this flight so we could relax. The realization that I was really going home finally began to hit me as I thought about the people back in Arkansas. Since I had not slept much the night before, this was a good opportunity for me to catch up on some sleep.

On our way back we had one refueling stop and then landed at an airfield in New Jersey where the Army picked me up and took me to Ft. Dix. This was the same place where I left the States to go overseas. When I arrived, they were expecting me, and I was processed right away, but had to spend the night there. The next morning I found that I was assigned to Santa Ana, California for six weeks of Rest and Relaxation. They issued me a Greyhound Bus ticket, with a delay en route to Little Rock, Arkansas, for two weeks. I was very happy about that delay because Little Rock was my hometown. This provided me with the opportunity to visit with family and friends that I had not seen in three years.

As soon as I could I got to a phone to call home to tell my folks when I would be there. I talked to my mother and sister and one of the first things they wanted to know was what food would I like for them to prepare for my homecoming. My first reaction to that question was, of course, a good old southern meal. It didn't take long for me to give them a response. I didn't even have to think about it. I asked: "How about country fried steak, whipped potatoes with lots of butter, homemade biscuits and gravy, corn, and black-eyed peas. And oh yes, if you happen to have any okra, you might fry up some of that also. For dessert, I would like apple pie with ice cream. Then I asked, "How about that for starters?" This was not an unusual meal for my mother for she was the type of cook who would cook everything in sight, just so there would be plenty to eat on the table. She never wanted anyone to go hungry. When I got there every item I requested on the phone was on the table. They didn't leave off one single dish of food I requested. It was wonderful.

It was good just to be back home, sleep in my own bed, and get caught up on what all had been going on while I was away. My Mom and Dad both seemed to be in good health and for that I was thankful. Many of my friends came by to see me and renew old acquaintances and we always promised to "stay in touch."

It seemed time was way too short for it was not long until I got back on the bus and headed for California. There on the same bus, I met Cpl. James Perkins who also was going to Santa Ana. He was from El Dorado, Arkansas, had just rotated back home and was being sent to the same base that I was for a little R&R.

Upon arriving in Santa Ana there was a Lieutenant who met the bus we were on and took us out to the base. He introduced himself and said that he would be our companion to show us around for the time we were out there. This base was a fine impressive layout. I didn't know the army had such excellent quarters as this anywhere. It was unlike anything I had seen in the service. Our accommodations could not have been any nicer, and we were treated like someone special. When the Sergeant at Ft. Dix sent us out here, he didn't tell me what we were getting into, but we sure got the VIP treatment when we arrived. There were only four of us assigned to this Lieutenant who was responsible for taking care of our every need. To start with while we were at this base we received a general physical examination, a trip to the dentist, issued new clothing. He even arranged our meals to see we got the food we liked. If we liked big thick steaks, we could have them as often as we wanted them. When our new clothes were issued, he saw to it they fit well and that all patches and decorations were on our uniform. The Army had issued one gold hash mark for each of the six months we served overseas. These were to be sewn on the left sleeve of my dress coat. When they finished sewing the seven hash marks on my coat they covered the entire area from my elbow to my wrist. Man was that ever impressive! Most people would have two or three gold hash marks on their left arm, but I had so many my uniform sleeve looked like a barber pole.

I had not been to a dentist in four years, because there were no dentists where I was, at least I never saw one. I had an appointment with a Navy dentist who examined my teeth and filled 11 teeth at one appointment. With no shots to block the pain I sat in that dental chair for over two hours and sweated a lot. I'm not sure who suffered the most, the dentist or me. It was hard on both of us. But this was necessary and I was glad to get it over as we had far better things to do.

One day while we were in California

our officer took us to the Santa Ana racetrack to see the horses run. On another day he took us to Beverly Hills to see the Stars' homes, the Hollywood Bowl for a concert that night and many other tourist sights. One thing we did that I really enjoyed was an entire day at one of the studios. He took us to other studios on a different day. The first day we were there, we were taken to the set of "Deception" where Betty Davis and Claude Rains were filming a scene. Next we went to watch Tyrone Power, Gene Tierney, and Anne Baxter making the film "The Razor's Edge." These were all interesting to see but the greatest thrill for me came when we went on the set of "Back To Bataan," featuring John Wayne and Anthony Quinn.

When the four of us arrived on this set John Wayne walked off the set as soon as he saw us and came over to greet us. We were introduced to the movie stars and the Duke sat down with us. Anthony Quinn excused himself by saying he needed to study his lines some more. John Wayne was glad to see service men and was very relaxed and friendly to us. He was a heavy smoker and the first thing he did was reach for a cigarette. I noticed this and at the same time I reached for my lighter to light his cigarette. He wanted to know about each of us, where we had been and what we did in the service. I told him about seeing one of his movies in Italy with a voice over where he was speaking in Italian. The Director, Edward Dmytryk, I could tell, was getting nervous about this interruption and with John Wayne's taking so long to visit with us. He said, "We need to get back into production." John Wayne completely ignored his remarks and continued to ask us questions. Finally in utter desperation the Director said, "Mr. Wayne, we must get back to work. Do you realize how much this delay is costing the studio with everyone just standing around?" John Wayne quickly answered the Director, "Do you realize how much these men have already paid in sacrifices just so we can be here making these films. Now, I will tell you when

we go back to work!" He turned to me and asked, "Now Sergeant, how about all those gold stripes on your left arm. Why don't you tell me what they mean and how you got them?" I of course complied with his request but we felt that we should leave as soon as we could do so.

The officer that had taken care of us so well here in Santa Ana, took us out to dinner that night in one of the fine supper clubs where we saw a very good entertaining show. I realized our time here was getting short and we would soon be leaving, so I had an opportunity to thank our host for taking such good care of us for the past six weeks.

"You know Lieutenant," I said, "When we came out here to Santa Ana, I thought we were just ordinary solders and everybody returning to the States would be treated just like us. I am beginning to think that is not so. I never thought of myself as a hero, but here we have been treated like we are someone special. You and the Army have really done a good job of making us feel welcome back home and making us feel we are special."

"You know Sergeant," the Lieutenant replied, "You men have been treated special, because the Army is proud of the job each of you has done for your country. You are right, all returning GI's will not get the same treatment you have had. To express our appreciation we have taken you around to show you off to the America that you have served so well. When John Wayne spoke of the sacrifices you men have made for your country, he very well spoke for all the people."

Each of us felt privileged to have been a part of this program, and expressed our thanks to the Lieutenant. The next day we were reassigned to a new outfit. The four of us were going in four different directions. I was assigned to McCook Air Force Base at McCook, Nebraska.

I thought how ironical that I would be going to Nebraska. That is the state that my friend Ray who was lost in the Ploesti raid, was from, and it was his parents that wanted me to come to Nebraska to visit them after the War. They wrote several times, almost insisting that I come. I thought about this opportunity and wondered if I should go see them. After weighing it over in my mind, I decided at this time it was too heavy for me to handle. I had already closed that door, and felt now was not the time to open it again. I needed time on my own to think through some things. Perhaps someday, but not now.

I was very happy to be at the McCook Air Base. The people there were friendly and glad to have me as part of their Photo Lab. It was quite a change from the pressure I felt with the 98th Bomb Group and this gave me time to readjust to living again, to be rehabilitated, to expand and to think. This place was exactly what I needed. The men here showed respect and were kind to me. They even cut me some slack, so that I could move at my own pace.

Some of my new friends invited me to get a three-day pass and go with them to Denver, Colorado. I was happy to do so for I had never been there. From our base, we took a 25-mile drive into North Platte where we left the car and caught the "Silver Bullet" into the Mile High City. It was only 260 miles and on this fast train it was a quick and pleasant trip. We did all the tourist things on a sight seeing trip around the city, and saw some beautiful sights. The thing that was most impressive to me was the people who lived there. They were warm and friendly, loved their city and loved having service men and women there. The people there were quick to show this attitude. Of all the places I have been I think the people of Denver were the most gracious people I have seen anywhere. Let me explain.

We were downtown waiting to go to a show we wanted to see, and passed a men's clothing store. I wanted to go in and look around and see what the prices were like. It had been a long time since I had bought any "civies." You could tell what was on my mind, because it wouldn't be long until I should be out of the Army. While I was just looking around, I saw a tie that I really liked and asked the clerk the price.

The store manager came over and said, "Just for you, it is free. If you like this tie take it as a gift from me in appreciation of your service to our country."

I said to him, "Sir, I thank you very much. The people of Denver are special people. They have bought me so many meals and coffee and now this tie. You are most generous, and I shall remember you for this kind gesture."

After a fun weekend of good food, excellent entertainment, and fellowship, we arrived back at base in McCook much refreshed and in good condition.

This was at a time when everyone loved servicemen. The general public just couldn't do enough for us. If I were in a restaurant eating, almost always someone would pick up my check and pay for my meal. Sometimes I never knew who it was so that I could thank them. Strangers would walk up to thank me for just being in the service, and then offer to buy me a cup of coffee, or do something for me. People were very generous and caring towards all service men and women. It was a good time to be in the Service, and it was especially good to be back home.

Now that the War was over in Europe, I began to hear about the 98th Bomb Group returning to the States. I even heard they might be coming to McCook Air Base. That would really be something neat if my old outfit caught up with me, but they never did.

While at McCook I had time to think about the years I put in with the 98th and the record this Bomb Group established as one of the most outstanding in the history of the Air Force. During the two years and eight months that the 98th was in the Mediterranean for a tour of duty, we flew missions without turning back a single time due to enemy fire. Since the Group began flying bombing missions out of the Middle East in 1942 until the last mission on April 15, 1945, there was a great turnover in both equipment and personnel, including those lost in action and those returning to the States. From Palestine, across North Africa and on into Italy this Bomb Group wrought great havoc to the Axis Forces in so many ways. They pounded the enemy's supply routes in the Mediterranean, flew support for Montgomery and the British forces all across the desert country. They made Air Force history in flying the first low level attack on the Ploesti oil fields on August 1, 1943. We flew from North African bases to deliver tremendous destruction to this vital source of oil for the Germans. From Italy the 98th continued to deliver devastating blows time after time to manufacturing plants deep in the Reich, at troop concentrations in the Balkans, and to keep bridges and communication lines cut through the Alps. In so many different ways the 98th Bomb Group denied the enemy the capacity to make war. In delivering this kind of destructive power to the enemy, the 98th had some losses too. We flew a total of 417 bombing missions, dropped 17,400 tons of explosives that were within the target area, with a loss of 108 B-24's. Over a thousand lives were lost or missing in action. From acts of heroism in the air, and on the ground, and from the many sacrifices by so many men with the attitude that we could not fail, this Bomb Group established a record unlike any other in the annals of world air history. I was proud to say that I was one of the members of the 98th Heavy Bomb Group. I was a Pyramidera.

Our group flew all B-24 Liberators, a powerful and a very versatile strategic World War II bomber. It was the workhorse of the Air Corps. The B-24 could carry more bombs and take them farther than any other plane of its time. There were over 18,000 of these B-24 planes built from 1939 through 1945. They were originally built by Consolidated's plants in San Diego, CA and Ft. Worth, Texas, but when America entered the war additional production was given to Ford Motor Co. at its Willow Run plant in Michigan and their plant in Dallas, Texas. In these plants and at the end of each hour a light would go on and a loud bell would ring as another B-24 rolled off the assembly line.

The B-24 plane saw duty in all theaters of the war including the ice and snow of the Aleutian Islands, the cutting winds, sand and heat of North Africa, and the heat and humidity of the tropics. In all locations, the plane adapted to the area and performed well in all of its tasks. Whatever job needed to be done this plane could do it. The success of the highly decorated 98th Bomb Group was due primarily to the high performance of the B-24 planes and the men who flew them.

The War in Europe was over and the Army Air Corps had decided to discharge me. For a college student who volunteered for one year of service in January of 1941, it was about time. That one-year stretched out to nearly five years. At last, I was glad to be going home. The Air Corps put me on a bus to Little Rock to be discharged there for that was where I entered the service.

Traveling on the bus alone, I had plenty of time to collect my thoughts and reflect on some of the things that had happened as well as what the future may hold for me. At a time like this there is a tendency to put aside the unpleasant things that have happened, try to forget them and to remember only the pleasant ones. I realized now it was possible for me to return home and complete my education. This was my number one priority, because that would help me get back into the swing of civilian life. I also recalled the day I left New York City on my way to war and the great desire I had to visit this city under different circumstances. I made a promise to myself that if I returned from this War I would visit New York City and to get the feel, the flavor and to experience the mysteries of this world renowned city. While the miles rolled by on the bus, I decided I would do this. After all I had visited Cairo and Rome and spent some quality time there, and now it is time I get to know New York City. At the end of summer and before school started I would go to New York City and stay a month just exploring this great city by myself. I could see all the shows on Broadway, ride the subways, eat at the Auto-mat, and do all the other things as I wanted to do them. This sounded like a workable plan to me, and I determined I would go for it.

Also, I thought about the Press conference that was held on the steps of Little Rock Central High School by members of the local and National Press in 1938. I remember so clearly the moment when a member of the press corps turned to me and asked, "Do you think the youth of America would fight in the event of War?" I wish we could be together again and hold a sequel to that press conference. I had some things I wanted to say to the members of the press. Thoughts were rolling through my mind, and they went something like this:

"Well America, you have your answer now. Your youth did fight the battles for you. And now, they are coming home, except they are no longer youth. We are men and women who no longer are willing to fight and die for you. We are tired of destroying, and now we are ready to build, and live for you. We want to build your streets and highways, your schools and libraries, your hospitals and medical centers, your churches and synagogues. We want to build on dreams that haven't even been dreamed yet. We want to explore space and discover new and greater opportunities. We want to respect our fellow man so that we can live together in peace and harmony. We want to build a more perfect America, where dreams can still come true. We have fought for America, now help us to be a part of this dream of ours, to build for America."

And surely, these are the thoughts, the dreams and the prayers of so many American GI's who have gone off to serve their Country and fight for the things in which they believed. We have come home. *Now* is the time to build on the things we fought for.

After I got out of the service, I spent my time at home bumming around and getting reacquainted with life. I had a great time visiting old friends and relatives. But things had changed. It was not as it had been. Some of my friends were not so fortunate and did not return from the war. Others had moved away or relocated somewhere else. But I spent the next three months finding myself and deciding what I wanted to do.

I followed the plan I had thought about on the bus trip home. I enrolled in Rochester Institute of Technology, Rochester, NY, dropped off most of my clothes and personal things and went on to New York City before school started. I checked into the YMCA and made my home there for the next six weeks. I lived on no time schedule and the only objective I had was to experience life in the big city to its fullest. I did just that. I saw all the Broadway shows, and many of the off-Broadway shows as well. I got around on the subways like a native, visited Wall Street, ate at the Auto-Mat, went to the top of the Empire State Building, and did it all. The intrigue of the city got to me on our way out to go overseas, and I made a vow to someday return and experience life there. Now I had done that and was ready to settle down to finish my education.

Since my college education was interrupted by the War, I was now ready to settle in for the next two years to finish what I had started. I went back to Rochester and got settled in for school. I met my future wife there, and after a year of dating we decided to get married. Our first year there, school was out on Friday and we were married the next morning. The church was full of students we were in school with; they stayed over to come to the wedding and wish us well. After another year, and graduation, we took off for Little Rock where we made our home.

My folks had a reception for us to introduce my new bride to relative and friends. It was a rather large crowd and one of those in attendanc was the Reverend Henry Adams, the father of my brother-in-law, Ken Adams. That night, he told my wife he had known me most of my life and he was sure that someday I would be a minister, and she would be a good minister's wife. Later that evening, as we were preparing for bed, she told me what Reverend Adams had said. I told my wife that it was the first I had heard of it. Besides, I said, Reverend Adams couldn't call me into the ministry; only God could do this and I hadn't heard anything from Him on this.

For the next 34 years, I continued to work in the church as a lay person in whatever capacity I was needed; I also worked in the secular community to earn a livelihood for my family. I was both successful and happy doing this. I still felt that I had lived a protected life, but I didn't understand exactly what I was supposed to be doing. It had been thirty-four years since Jesus appeared to me on that airplane, and I was still wondering if I would ever know. In the meantime, I would keep on doing what I could for the church. At the age of 56, I decided to sell the business I owned and make a career change. Everyone was asking what I was going to do. I told them I didn't know (which was true), that I was too young to retire and too old to worry about it. All that I knew for sure was that we were moving back to Little Rock where I started out. I was going back to find my roots.

After moving back to Little Rock, we attended First United Methodist Church, where Dr. Alvin Murray was the senior Pastor. One Sunday he preached a sermon on "We are called into a ministry of service"; we just needed to discover what our ministry was. I thought about this all weekend long. Monday I called Dr. Murray and made an appointment to discuss this with him. We talked at length about a ministry for me, had prayer, and went in to talk further with Bishop Ken Hicks. When I left his office that day I had been enrolled in a "License to Preach School" and was on my way into the Ministry. I had discussed this with my wife and she was very supportive. That evening after supper we discussed the events of the day and felt we had made the right decision. We both felt very good about what had happened.

I have always been a very sound sleeper and would very seldom dream. If I did dream at all, the next morning I wouldn't remember anything about it. That night was no exception. The next morning, we awakened had breakfast, read our devotion for the day, and I went upstairs to finish dressing. My wife stayed down at the breakfast table to read the Bible further. As I began collecting the papers on top of the chest of drawers, I saw a note that read "Hebrews 13:20-21." I looked at it and wondered when I might have written that down. I had not been reading in Hebrews.

All of a sudden, I remembered tossing and tumbling in the night at the sound of a voice repeating, "Hebrews 13:20-21."

I said, "Okay, okay, I will remember it if you will just let me go back to sleep."

Then I heard a loud voice say, "NO YOU WON'T. NOW GET UP AND WRITE IT DOWN."

I quickly jumped out of bed, got a pen and paper, wrote it down and went back to bed without turning on a light. I promptly went back to sleep and forgot about the incident in the night.

This all came back to me as I read what I had written. I yelled downstairs, "Honey, look up Hebrews 13:20-21 and read it." At the same time, I hurriedly reached for the New Testament that I carried in my coat pocket, thinking "Oh Lord, let there be a thirteenth chapter." (I knew that Hebrews was a rather short book, but I didn't remember how many chapters were in the book. I had not been reading in that book for some time.) I turned to it and began to read:

*Now may the God of peace, who brought back from the dead our Lord Jesus, the great shepherd of the sheep, by the blood of the eternal covenant, equip you with everything good that you may do his will, working in you that which is pleasing in his sight, through Jesus Christ; to whom be glory for ever and ever. Amen. (RSV)*

I met my wife on the stairway; we embraced, and I said that this was confirmation that I had made the right commitment. Now I knew what Jesus had meant when he said, "In time you will know, and you will know that you know." I had waited 34 years for this word, and now I knew with an assurance what I was to do. This scripture was telling me that He would create in me the things that were pleasing to Him, and He would equip me with everything good to do His will. With this confirmation, I could enter the ministry with confidence. Once again, just as I had experienced in my personal encounter with Jesus 34 years ago, I felt that I was living a secured life. Doors were opened, paths were cleared, and I entered the ministry with an assurance that I was following God's plan for my life. God has been good to us. We have been blessed and have had a rewarding and satisfying ministry.

Once I made commitment to go into full-time ministry and was ordained, I was able to help more individuals than ever before. I have lived a life of service to others. People who recognize me as a minister are more receptive to the word that I bring. They have found a healing ministry and they find peace and comfort in their lives.

If we respond to Christ only from a sense of duty, our lives become distorted and out of focus. It is in reaching out to others, that extra commitment that we make, that produces a full and gratifying life. "If you keep my commandments, you shall abide in my love.... These things have I spoken unto you, that my joy might remain in you, and that your joy might be full" (John 15:10-11).

This life is not a dress rehearsal, where if you don't get it right the first time, you can do it over again. The great gift of life itself must never be taken for granted. It is the real thing. We must accept it with the awesome responsibility that goes along with it. If we do this, it can be the most richly rewarding experience we can have. This has been so far for me.

And finally, for me, 'out of the desert' came an unshakable faith in God. It was here that I had a life-changing experience with Christ. It was here that I was tried, tested, and found to be faithful. No matter how difficult things were, I had found a friend in Jesus, and He saw me through. He nurtured me, and provided me with the strength I needed to make it. He took care of my every need. The strength of my faith came from His word: "And remember, I am with you always, to the end of the age" (Matthew 28:20). This was the attitude with which I came through the war; it was the attitude that brought me safely home to live another day.

Thanks be to God.

# Chapter Ten

## Air Offensive/Awards & Honors

### 98th Campaigns

Egypt, Libya, South Europe, Tunisia, Sicily, Naples, Foggia, Anzio, Rome-Arno, Southern France, and Rumania.

417 Total missions flown from August 15, 1942 to April 11, 1945.

17,336 Total bomb tonnage dropped in or near the target area.

### Awards & Honors

*President's Distinguished Unit Citation*
Awarded to 98th Bombardment Group on August 17, 1942 for support of British Forces in North Africa and Sicily.

*President's Distinguished Unit Citation*
Awarded to 98th Bombardment Group on August 1, 1943 for the August 1, 1943 low level attack on Ploesti, Rumania.

*Medal of Honor*
Bestowed by the President in the name of Congress for deeds of surpassing valor, and devotion far above the call of duty:

Colonel John R. Kane
Major John J. Jerstad
Second Lt. Lloyd H. Hughes
Lt. Colonel Addison E. Baker

*B-24 coming in low over Ploesti target.*

*Bombs dropping on Maritza Air Base in Italy.*

# CHAPTER ELEVEN

*Combat Crews that flew August 1st Mission over Ploesti*

## Plane 662-Black Magic
Dwight D. Patch
John C. Parks
Phillip G. Papish
William J. Reynolds
Ellis J. Bonorden
Richard F. Fulcher
Kenneth W. Parish
John A. Ditullis
Joseph J. McHune
Emiel F. DeBaets

## Plane 195-Little Joe
Wesley N. Pettigrew
William J. Bergan
John F. Staehle
Oeva A. Weijanen
Lawrence L. Scholl
Lowell A. Folks
James O. Leming
Jack L. Swafford
Howard E. Leon

## Plane 758-Lil Jughaid
Robert G. Nicholson
Harry J. Baker
Oscie K. Parker
Boyden Supiano
Joseph G. Redfield
Walter Cybulski
Clarence A. Laidlaw
Manual R. Rangel
Roy B. McCracken
Donald J. Osburn

## Plane 782-Boilmaker II
Theodore E. Helin
Charles E. Smith
Albert M. Arnson
Maynard G. Hubbard
Harry C. Opp
Arthur W. White
Delbert R. Warner
Raymond C. Walaczaka

Peter C. Passalacqua
William M. Zeger

## Plane 195F-Penelope
Martin A. Speiser
Daniel H. Walls
Robert C. Stephens
William E. Wright
Donald C. Yocum
Warren A. Orr
Larry M. McCabe
Elbert M. Payne
Crosby M. Smith
Roy S. Foy

## Plane 815-Daisy Mae
Lewis N. Ellis
Callistus E. Fager
Julius K. Klenkbell
Guido Gioano
Arthur T. Waugh
Nicholas C. Hunt
Carl A. Alfredson
Owen J. Coldiron
Blase B. Dillman
James W. Ayers

## Plane 817-The Stinger
Andrew W. Opsata
Louis R. Quaglino
Howard J. LaLonde
Donald N. DiCoscoe
Charles P. Quinlan
Paul A. Nicholson
William G. Pimm
John H. Gormey
John M. Oakes

## Plane 836-Lil De-icer
James L. Merrick
howard A. Schaufele
Elton R. Stutling
Jerome S. Victor
Hugh D. Nelson

Robert J. Cooper
Willaim D. Isaacson
Edward L. Haydeck
Eugene M. Gambrell
Eugene P. Tomlinson

### Plane 023-Fertile Myrtle
Herbert I. Scingler Jr.
Lyle A. Spencer
Albert V. Freeman Jr.
Jack W. Kaboth
Robert H. Johnson
George W. Fulfer
Edward F. McSweeny
Ernest F. Knutson
Seth E. Ely
John W. Morgan

### Plane 322-The Cornhusker
Ned Mcarthy
Clyde E. Miller
Joseph F. Moore
Robert L. Price
Claude C. Roberts
Dewey O. Jackson
Glenn D. Darr
Glenn L. Koontz

*Staff Sgt. Eugene Gambrell prepares to mount an aerial camera in a B-24.*

Stephen Wargo
Clayton E. Ballard

### Plane 520-Yen-tu
Edward Mcguire
James H. Marrah
Max H. Schweitz
Russell H. Godde
Turner Y. Johnson
Harry G. Konecny
Moses F. Tate
James R. Waltman
Robert Rans
Clark S. Fitzpatrick

### Plane 246-Semper Felix
August W. Sulflow
Leroy P. Miller
Phillip P. Miller
Anel B. Shay Jr.
Leornard E. Reger
William E. Treichler
Vincent L. Politte
Norman L. Meyer
Henry A. Gallas
Jack O. Samson

### Plane 774-Chief
Thomas P. Fravega
James R. Morgan
Edward Strickbine
Oscar F. McWhirter
Anthony T. Fravega
James Gabehart
Bronislaus C. Pitak
George H. Parramore
Eric Hurt

### Plane 663-Maternity Ward
John V. Ward
Andrew L. Anderson
Beverly S. Huntly
Henry C. Crump Jr
James J. Toth
Leon D. Pemberton
Robert E. Long
Kenneth L. Turner
Harold W. Scott
William J. Fay

### Plane 991-Kate Smith
James A. Deeds
Clifton C. Foster
Francis V. Montemurro

*Little Chief, Big Dog.*

Theodore F. Scarborough
Frank E. Leising
James M. Howie
Donald A. Pimlott
Joseph T. Cupina
John W. Potts
Adolph Oleenik

### Plane 886-Lil Joe
Lindley P. Hussey
Donald Jenkins
Phillip E. Nelson
Allan E. Peterson
Lloyd T. Fowler
Edmond T. Terry
Roy C. Carney
Raymond A. Heisner
James E. Turner

### Plane 102-Old Baldy
John J. Dore Jr.
John B. Stallings
Worthington A. Franks
Joseph E. Finneran
Frank A. Norris
Wesley L. Jones
Stanley R. Packer
Joseph R. Iosco
Ray L. Gleason

### Plane 661-Black Jack
Delbert H. Hahn
John W. Viewers
Nathat Nowak
Raymond F. Vengelen
Robert L. Baird
James E. Creighton

Isadore I. Klein
Leslie J. Foster
John H. Chapman
Curtis Washburn

### Plane 766-Chugalug
Leroy B. Morgan
Hosey W. Rich
Victor H. Larson
Chaplin J. Watkins
James Van Ness
Robert L. Tipton
Norbert I. Petrie
Norman I. Cupp
Forrest D. Hudley

### Plane 402-The Sandman
Robert W. Sternfels
Barney Jackson
Anthony W. Flesch
David A. Polachek
William W. Stout
Frank Just
John T. Weston
Raymond E. Stewart
Harry Rifkin
Merle B. Bolen

### Plane 316-Snake Eyes
Hilary M. Blevins
James E. O'Grady
William Toles
Harold G. Moore
Harold G. Knotts
Mark B. Weber
John Probst
Norris F. Dietrich
Barner F. Clemens
Ernest V. Martin

### Plane 312-Aire Lobe
John B. Thomas
David M. Lewis
Robert D. Nash
George McCandless
Elijah D. Johnson
Eugene E. Gough
Richard G. Salisbury
George E. Davies
William A. Kneisl
Eldon L. Carter

### Plane 973-Battleaxe
Charles A. Salyer

Nicholas G. Cilli
James D. Grover
Walter E. Mackey
Roy D. Hammond
George Aguayo
John A. Verbitski
Ralph C. Wessel
Harold L. Pace

### Plane O26- Baby
Francis E. Weisler
Francis A. McClellan
Paul A. Warrenfeltz
Joseph L. Nagy
Wayne H. Rutledge
Paul A. Joyce
Arlo D. Beem
Malcomb D. Smith
Adney J. Harmon
Chester L. Dell

### Plane 198 Vulgar Virgin
Wallace C. Taylor
Paul W. Packer
Jack C. Wood
Robert N. Austin
Gerald E. Rabb
Alfred F. Turgeon
Donald R. Dunchene
Arthur B. VanKleek
Ralph M. Robbins
Louis Kaiser

### Plane 768-Kickapoo
Robert J. Nespor
John C. Riley
Russel W. Polivka
Thurman L. Ward
Vaund D. Wenrich
Armand R. Massart
Eugene R. Garner
George W. Lawlor
John P. D'Armour
Edwin G. Sliwa

### Plane 803-Rosie Wrecked 'Em
Herbert W. Arnes
Hamlin Kitteridge
Joe A. Ellison
Thomas J. Shyrock
Charles A. Lawson
Warren O. Kidder
Martin H. Malony
Ancil C. Holman

*B-24 aircraft en route to target over Italy.*

Harry L. Kaminski
Joseph J. Fasano

### Plane 040-Big Operator
Hoover Edwards
Homer S. McCullom
John R. Fontenrose
Albert J. Mickish
Bernard M. Boullioum
Gene J. Colley
Dale R. Colley
Raymond Waters
John S. Alesaukas
Vernon C. Utter

### Plane 819-Raunchy
Samuel R. Neeley
Herman H. Henslee
Joel I. Corn
Eugene L. Rodgen
Earl T. Edelen
Robert P. Schultz
William W. Schiffmacher
Charles P. Geers
Nick A. Allen
Alfred D. Cason

### Plane 626-Rowdy II
Allen B. Gaston
Daniel J. Mullins
Hubert Skembare
George C. Roman
James T. McKinley
Bernard E. Arciero
Thomas Kincaid
Leslie A. Johnson
Harry J. Flanagan
Torsten W. Ahlbeck

## Plane 795-The Squaw
Royden L. LeBrecht
Clinton H. Killian
Grover A. Zink
James H. Faulkner
Harold F. Weir
John A. Givens
John R. Reilly
John Guani
Paul E. Davidson
John S. Potvin

## Plane 733-Skipper
George Colchagoff
Lorin P. Kinkaid
Robert K. Lamberts
James W. Kendall
William N. Wellons
Robert J. Ruhl
William Y. Cox Jr.
Thomas Hubbard
Myron H. Kinsley
Charles L. Peterson

## Plane 208-Sad Sack
William Banks
Carl Root
Theodore Stewart
Joseph Souza
Earl Rice
Durward Carbury
Wilson Cane
Floyd Pleasant
Walter Golec
Henry Richotte

## Plane 311-Hadley's Harem
Gilbert B. Hadley
James R. Lindsey
Harold Tabacoff
Leon M. Storms
Russell B. Page
William F. Leonard
Frank Nemoth
Pershing W. Waples
Leroy Newton
Christopher N. Holweger

## Plane 825-Hail Columbia
John R. Kane
John S. Young
Norman W. Whalen
Harold F. Korger
Fredrick A. Laerd
William Leo
Raymond B. Hubbard
Joseph J. Labranche
Harvey L. Treace
Neville C. Benson

## Plane 776-Jersey Jackass
John J. McGraw
Charles D. Cavit
Robert J. Seniff
George F. Giblin
Warren T. Townley
Albert H. Wilmes
William P. Sheridan
James A. Utely
John R. Ross

## Plane 840-The Witch
Julian T. Darlington
Daryl P. Epp
Joseph N. Quigley
Major R. Gillette
Lloyd W. Brisbi
Dale G. Hulsey
Ned A. Howard
Anthony J. Rauba
Joseph J. Turley
Walter D. Hardiek

## Plane 767-Shanghai Lil
Carl S. Looker
George L. Clark
Hereau E. Stoddard
David D. Meese
Gregory B. Crock
Milton P. Remley

*Crew of The Squaw pose in front of their plane.*

Louis R. Strong
Joseph A. Henaire
Fred M. Drugger Jr.
Julius A. Baca

## Plane 795-Sneezy
Donald G. Johnson
John P. Foley
Lloyd E. Hooloway Jr.
Charles K. Warren
Thomas O'Leary
Louis J. McNamara
Robert C. Lindsey
John Onysczak
William A. Cornut
Yves J. Gouin

## Plane 655-Four Eyes
Lawrence Hadock
John P. Draft
William R. DeBusk
Peter A. Timpo
Lawrence E. Reitz
Zelwood A. Gravlin
Robert C. Elliot
Eugene O. O'Mara
James B. Hale

## Plane 082-Scarlet O'hara
Thomas W. Bennett
Kenneth C. Gray
Raymond C. Meyers
Vernon L. Miller
John P. Jones
Jules E. Livingstone
Samuel W. Greelee
Nelson C. Scott
Robert H. Nettleton

## Plane 896-Nightmare
Julian N. Bleyer
David S. Watt
Charles M. Parker
Jack H. Rumsey
Russel E. Burton
Bobby E. McCowa
Robert E. Looke
Frank B. Kozak
Francis Beauregard
Jerry Joswick

## Plane 921-Northern Star
Glen W. Underwood
Jean W. Gambrill
Edward Rothkrug
Robert J. Judy
Clement S. Badeau
Ernest E. Sestina
Francis S. Beatty
Henry B. English
Rexford H. Rhodes

## Plane 313-Boots
Lawrence E. Murphy
Gilbert H. Kyer
Joseph T. Rotundo
William S. Havens
John B. DeLoire
William L. Popham
Edward J. Amand
Donald P. Sowers
Harold N. Yost
Thomas J. McGrath

## Plane 364-Prince Charming
James A. Gunn III
Richard C. Williamson

*Entire photo crew with Alex in Italy.*

*One of the large aerial cameras in use on missions.*

Bernard E. Leimbach
Robert E. Courtney
Edwin L. Turner
Delmar E. Smith
Stanley M. Horine Jr.
Fredrick L. Stukey Jr.
Charles R. Ruark

### Plane 007-Margie
Clarence W. Gooden
Jerome D. Savaria
Ralph F. Parkins Jr.
William H. McNeil
Michael J. Trick
Theodore C. Beaudy
George H. Kaylor
Harry G. Deem
Alexander M. Cochrane
Roland B. Cox

### Plane 197-Tagalong
Ralph V. Hinch
Charles C. Barbour
James G. Taylor
Stanley J. Samoski
Robert F. Mead
Paul F. Eschelman
Harry G. Baughn
Donald G. Wright
Robert C. Coleman
Delmar M. Schweigert

### Plane 716-Yours Truly
Elmer R. Rodenberg
Doyle Hicks
Harold L. Polinsky
George B. Hammond
Leslie W. Martin
George K. Holroyd
Robert J. Clark
James W. Morgan
William C. Eggleston
Marcellus L. Oberste

*B-24 over Greece.*

*Inspecting and interpreting bomb-strike photos. Author, Capt. Brown, and Lt. Fagan.*

| Plane Number | Pilot | MISSION BOARD<br>Take off | Landed/Remarks |
|---|---|---|---|
| **SECTION A** | | | |
| 662 | Patch | 07:33 | 21:05 |
| 195 | Pettigrew | 07:55 | 21:11 |
| 758 | Nichelson | 07:30 | 21:10 |
| 782 | Helin | 07:34 | Down over target |
| 195 | Speiser | 07:57 | 21:04 |
| 815 | Ellis | 07:31 | 21:15 |
| 817 | Opsata | 07:36 | 21:08 |
| 836 | Mervier | 07:29 | 21:05 |
| 023 | Shingler | 07:13 | 21:07 |
| 322 | McCarthy | 07:32 | Down near target |
| 520 | McGuire | 07:41 | Missing |
| 246 | Sulflow | 07:11 | Missing |
| 774 | Fraveca | 07:16 | Down at Sicily |
| 663 | Ward | 07:17 | Missing |
| 991 | Deeds | 07:17 | Missing |
| 886 | Hussey | 07:22 | Missing |
| 102 | Dore | 07:24 | Missing |
| 661 | Hahn | 07:12 | Down at Sicily |
| 766 | Morgan | 07:23 | 21:11 |
| 402 | Sternfels | 07:26 | Landed at Cyprus |
| 316 | Blevins | 07:19 | Landed at Sicily |
| 312 | Thomas | 07:27 | Missing |
| 973 | Sayler | 07:27 | 21:04 |
| | | | |
| **SECTION B** | | | |
| 782 | Helin | 07:21 | Missing |
| 026 | Weisler | 07:10 | Landed at Cyprus |
| 198 | Taylor | 07:09 | Missing |
| 768 | Nespor | 07:23 | 08:00 Crashed at Lete |
| 803 | Arnes | 07:15 | Landed at Malta |
| 040 | Edwards | 06:45 | Lost- Search |
| 814 | Neely | 07:20 | Missing |
| 795 | LeBrecht | 07:22 | Landed at Cyprus |
| 723 | Colchagoff | 06:45 | Lost- Search |
| 208 | Barnes | 07:19 | Landed at Cyprus |
| 311 | Hadley | 07:16 | Landed at Cyprus |
| 825 | Kane | 07:14 | 17:34 Crashed Cyprus |
| 776 | McGraw | 07:12 | Missing |
| 840 | Darlington | 07:33 | Down over target |
| 767 | Looker | 07:28 | Crashed at Sicily |
| 761 | Johnson | 07:34 | 21:14 |
| 655 | Hadock | 07:31 | Down over target |
| 082 | Bennett | 07:30 | Missing- Search |
| 896 | Bleyer | 07:25 | 21:05 |
| 921 | Underwood | 07:27 | 21:10 |
| 313 | Murphy | 07:30 | Missing- Search |
| 364 | Gunn | 07:32 | Down over target |
| 620 | Haverty | 07:13 | Engine feathered, left out |
| 007 | Gooden | 07:29 | Down over target |

110

# BIBLIOGRAPHY

*Books:*

Churchill, Winston, *Memoirs of the Second World War,* Houghton Mifflin Company, Boston, 1959.

Coffey, Frank, *Always Home: 50 Years of the USO,* Brassey's (US), Inc. McLean, Virginia, 1991.

Collier, Richard, *The War in the Desert,* Time-Life Books, Alexander, Virginia, 1977.

Dugan, James and Stewart, Carroll, *PLOESTI: The Great Ground-Air Battle of 1 August, 1943,* Random House, New York, 1962.

Faisg, Hank, Jr., *Force For Freedom-The Legacy of the 98th,* Turner Publishing Company, Paducah, Ky, 1990.

Gillies, Frederick W., *The Story of a Squadron, 154 Observation, Reconnaissance Squadron, 1946.*

Hill, Michael, *Black Sunday: PLOESTI,* Schiffer Publishing Ltd., Atglen, PA. 1993.

Johnsen, Frederick A., *Warbird History B-24 Liberator,* Motorbooks International, Osceola, WI, 1993.

Mauldin, Bill, *UP FRONT,* W. W. Norton Company, 1995.

Pyle, Ernie, *Here is Your War,* Henry Holt Company, 1944.

Sulzberger, C.L., *Picture History of World War II,* American Heritage Publishing Company Inc., 1996.

*Articles & Others*

"Ploesti Raid", Life Magazine, August 30, 1943.

Young, John S., Over the Target, Air Force Magazine, November 1943.

Compton's Encyclopedia, 99, Mindscape, Novato, CA, 1999.

Encarta 97 Encyclopedia, Microsoft Corporation, Redmond, WA, 1997.